Research for Policy

Studies by the Netherlands Council for Government Policy

The Netherlands Scientific Council for Government Policy (WRR) is an independent strategic advisory body for government policy in the Netherlands. It advises the Dutch government and Parliament on long-term strategic issues that are of great importance for society. The WRR provides science-based advice aimed at: opening up new perspectives and directions, changing problem definitions, setting new policy goals, investigating new resources for problem-solving and enriching the public debate.

The studies of the WRR do not focus on one particular policy area, but on cross-cutting issues that affect future policymaking in multiple domains. A long-term perspective complements the day-to-day policymaking, which often concentrates on issues that dominate today's policy agenda.

The WRR consists of a Council and an academic staff who work closely together in multidisciplinary project teams. Council members are appointed by the King, and hold academic chairs at universities, currently in fields as diverse as economics, sociology, law, public administration and governance, health and water management. The WRR determines its own work programme, as well as the content of its publications. All its work is externally reviewed before publication.

The Research for Policy Series

In this series, we publish internationally relevant studies of the Netherlands Scientific Council for Government Policy. Many of the cross-cutting issues that affect Dutch policymaking, also challenge other Western countries or international bodies. By publishing these studies in this international open access scientific series, we hope that our analyses and insights can contribute to the policy debate in other countries.

About the Editors

Corien Prins is Chair of the WRR and Professor of Law and Information Technology at Tilburg Law School (Tilburg University).

Frans Brom is Council secretary and director of the WRR office. He also is Professor of Normativity of Scientific Policy Advice at the Ethics Institute of Utrecht University.

More information about this series at http://www.springer.com/series/16390

Ernst Hirsch Ballin • Huub Dijstelbloem
Peter de Goede
Editors

Security in an Interconnected World

A Strategic Vision for Defence Policy

WRR
THE NETHERLANDS SCIENTIFIC COUNCIL FOR GOVERNMENT POLICY

Springer Open

Editors
Ernst Hirsch Ballin
WRR, Scientific Council for Government Policy
The Hague, Zuid-Holland, The Netherlands

Huub Dijstelbloem
WRR, Scientific Council for Government Policy
The Hague, Zuid-Holland, The Netherlands

Peter de Goede
WRR, Scientific Council for Government Policy
The Hague, Zuid-Holland, The Netherlands

ISSN 2662-3684 ISSN 2662-3692 (electronic)
Research for Policy
ISBN 978-3-030-37605-5 ISBN 978-3-030-37606-2 (eBook)
https://doi.org/10.1007/978-3-030-37606-2

This Springer imprint is published by the registered company Springer Nature Switzerland AG
The registered company address is: Gewerbestrasse 11, 6330 Cham, Switzerland

Preface

We are pleased to present the book *Security in an Interconnected World: A Strategic Vision for Defence Policy*. The book is a translation and adaption of the Dutch report 'Veiligheid in een wereld van verbindingen. Een strategische visie op het defensiebeleid', published by the Netherlands Scientific Council for Government Policy (WRR). The initial report was presented to the Prime Minister and the Minister of Defence on 10 May 2017. This book contains minor revisions and updated additions to text, most notably with regard to the increased budget of the Ministry of Defence and to the developments concerning European cooperation.

This book wishes to stimulate the political and public debate about the priorities and capabilities of an intelligent, future-proof security and defence policy for the Netherlands. The reference point for the framework of the actions proposed in this book is the armed forces, but in the editors' view, development cooperation and foreign policy also need to be strengthened if the Netherlands is to be strategically equipped for the geopolitical situation. This book investigates options for a future-proof international security policy and focuses on defence policy and the position of the armed forces in light of the changed geopolitical environment. The editors recommend developing the future armed forces on the basis of an integrated security strategy that encompasses both internal and external securities. National and international securities are increasingly interconnected. For the purposes of this strategy, the book recommends the establishment of a Netherlands Security Council and a Security Planning and Research Agency. NATO and the EU would remain the principal frameworks for defence policy. There is a need to tighten the focus of additional investments in the armed forces. In addition to compensating for earlier cutbacks, the starting point for investments should be the existing cooperation relationships with partners.

In the official response on behalf of the Council of Ministers to the 2017 report,[1] the Minister of Defence emphasises the Cabinet acknowledges the importance of the report and its analysis. Many of the recommendations were fully agreed upon,

[1] Parliamentary Documents II 2017–2018, 33 763, no. 141, 28 March 2018.

whilst other recommendations were welcomed. The response was more reluctant with regard to the more institutional recommendations, i.e. the establishment of a Netherlands Security Council and a Security Planning and Research Agency, and expressed some reservations about the recommended emphasis on specialisation. The Cabinet did follow up on this line of recommendations by publishing the 2018 Defence Note (Defensienota) and Integrated Foreign and Security Strategy (Geïntegreerde Buitenland en Veiligheidsstrategie, GBVS) together and composing them in a coherent way.

Since its publication in 2017, the WRR report 'Veiligheid in een wereld van verbindingen. Een strategische visie op het defensiebeleid' has stimulated the public and political debate. The report appears to be an ongoing source of inspiration to reflect on the strategic issues concerning defence policies in an interconnected world. The institutional recommendations continue to affect this debate as they encourage the reflection on the design of strategic defence and security policies. For that reason, this book presents the ideas of the original report to an international audience.

The book is the product of an extensive process of research, consultation and analysis. Ernst Hirsch Ballin, Huub Dijstelbloem and Peter de Goede are the editors of this publication. Together with Casper de Vries, they formed a project group, chaired by member of the Council Ernst Hirsch Ballin and coordinated by Huub Dijstelbloem. Important contributions were made by two former members of the project group, Wendy Asbeek Brusse, Former Director of the WRR, and Auke Venema, Former Coordinator of the project. Other previous members of the project include F.S.L. Schouten, W. Sediq and M. Verwijk. This publication presented by the aforementioned editors makes grateful use of the work of all people involved in the making of the original Dutch report.

In addition to studying the academic literature, all project members conducted numerous interviews with experts, policy-makers and stakeholders. We are very grateful for their time and effort. Their names are listed at the end of the book. Special thanks go to the experts who were prepared to read and comment at length on an earlier version of this book: Dr. Margriet Drent (Clingendael Institute), Prof. Dr. Paul Ducheine (University of Amsterdam and Netherlands Defence Academy), M. Kwast MA (Advisory Council on International Affairs), Dr. S. Reyn (Ministry of Defence), Professor J.G. de Hoop Scheffer (Advisory Council on International Affairs), Monica Sie MA (Clingendael), Colonel Wagner (Ministry of Defence), Professor Rob de Wijk (HCSS) and Dick Zandee MA (Clingendael Institute).

The Hague, The Netherlands

Ernst Hirsch Ballin
Huub Dijstelbloem
Peter de Goede

Contents

1 **Shifts in the Security Environment** . . . 1
Ernst Hirsch Ballin, Huub Dijstelbloem, and Peter de Goede

2 **The Extension of the Concept of Security** . . . 13
Ernst Hirsch Ballin, Huub Dijstelbloem, and Peter de Goede

3 **Human Security and Flow Security in Dutch Security Policy** . . . 41
Ernst Hirsch Ballin, Huub Dijstelbloem, and Peter de Goede

4 **The Netherlands and the Extended Concept of Security: The Rise of Security Strategies** . . . 65
Ernst Hirsch Ballin, Huub Dijstelbloem, and Peter de Goede

5 **Defence Policy in a Changed Security Environment** . . . 87
Ernst Hirsch Ballin, Huub Dijstelbloem, and Peter de Goede

6 **Strategic Strengthening of the Armed Forces** . . . 119
Ernst Hirsch Ballin, Huub Dijstelbloem, and Peter de Goede

7 **Interconnected Security: Conclusions and Recommendations** . . . 145
Ernst Hirsch Ballin, Huub Dijstelbloem, and Peter de Goede

Appendix I: List of Persons Consulted . . . 163

Appendix II: List of Acronyms and Abbreviations . . . 167

Bibliography . . . 171

About the Authors

Ernst Hirsch Ballin is Distinguished University Professor at Tilburg University and, as Professor of Human Rights Law at the University of Amsterdam, President of the Asser Institute for International and European Law in The Hague. He was a Member of the Scientific Council for Government Policy (WRR, 2014–2019) and was subsequently appointed as Consultative Member of the council. He is also a Member of the Royal Netherlands Academy of Arts and Sciences (KNAW) and of the Advisory Council on International Affairs (AIV). In addition, he was the Netherlands Minister of Justice (1989–1994, 2006–2010) and represented as such the Netherlands in the Council of the European Union for Justice and Home Affairs and was also a Member of Parliament (1994–1995 House of Representatives, 1995–2000 Senate), followed by his appointment to the Council of State (2000–2006). In 1999–2000, he was a Member of the Convention charged with drawing up the EU Charter of Fundamental Rights.

Huub Dijstelbloem is Professor of Philosophy of Science and Politics at the University of Amsterdam (UvA) and Senior Researcher and Project Leader at the Scientific Council for Government Policy in The Hague (WRR). As a Visiting Scholar, he was affiliated to the University of California San Diego (2014) and to the University of Maputo's Center for Policy Analysis (2010). He is involved in public debates about science, technology, and democracy and is one of the initiators of *Science in Transition*. His research is concerned with border control technologies, security policies, and migration issues and has been published in various books, edited volumes, and journals including *Nature, Geopolitics, International Political Sociology, Security Dialogue, Journal of Borderland Studies* and *European Journal of Social Theory.*

Peter de Goede is Senior Researcher at the Scientific Council for Government Policy (Wetenschappelijke Raad voor het Regeringsbeleid) in the Netherlands. He studied (comparative) political science at Radboud University in Nijmegen and worked at the same university as Assistant Professor in the Department of Public Administration. He obtained his doctorate at Leiden University with a thesis providing a comparative history of 80 years of public broadcasting policy in the Netherlands. He has also worked as a Senior Advisor at the Council for Public Administration (Raad voor het Openbaar Bestuur).

Chapter 1
Shifts in the Security Environment

Ernst Hirsch Ballin, Huub Dijstelbloem, and Peter de Goede

1.1 The Netherlands in the Shifting Geopolitical Force Field

The Netherlands' security and defence policy is entering a crucial phase. Since the end of the Cold War and the ensuing apparent supremacy of American hegemony, the geopolitical force field has changed radically. The last decade has seen the emergence of a multipolar world which is creating increasing tensions that directly affect the Netherlands. This new order calls for a strategic analysis of the security environment on the basis of which defence policy, and more specifically the role of the armed forces, can be formulated.

The Dutch armed forces have responded to the shifts occurring on the world stage and in the process have been highly active in a wide variety of operations. Since 1990 the armed forces have been transformed into an expeditionary organisation. Conscription has been suspended and professional Dutch soldiers have taken part in numerous international missions – approximately 50 since 1990. The structure of the slimmed-down Dutch armed forces was geared to these types of operations – with a fairly wide range of military capabilities, but without the resources to carry them out for longer than a few years. Now, however, the capacity of the armed forces is under severe pressure. It is time to re-assess the strategic position of the defence forces in the medium term.

The capacity of the branches of the armed forces has been eroded by the so-called 'peace dividend', which has been repeatedly cashed in since the fall of the Berlin Wall, and the many rounds of cutbacks that have been made. In 2016, the Netherlands Court of Audit [*De Algemene Rekenkamer*] found that "for years the material readiness of the armed forces has been below the level regarded by the Ministry of Defence as necessary to meet the objectives for deployability formu-

E. Hirsch Ballin (✉) · H. Dijstelbloem · P. de Goede
WRR, Scientific Council for Government Policy, The Hague, Zuid-Holland, The Netherlands
e-mail: ballin@tilburguniversity.edu; h.o.dijstelbloem@uva.nl; peter.goede@minbzk.nl

E. Hirsch Ballin et al. (eds.), *Security in an Interconnected World*,
Research for Policy, https://doi.org/10.1007/978-3-030-37606-2_1

lated by the Minister of Defence in the budget. We have already observed in some years that the Ministry of Defence is putting a lot of pressure on itself due to the imbalance between ambitions, resources and organisation."

In the meantime, the tasks the armed forces have to carry out have not diminished. The assignments that the armed forces must be able to perform to protect the territory and the inhabitants, to safeguard the international legal order and to support civil authorities are not going to subside in the near future. The nature of the threats has also changed. Increasingly, cyber attacks, threats to vital infrastructure and the use of disinformation (as created by Russia with respect to the investigation into the Flight MH17 disaster) are setting the tone.

The Netherlands does not operate alone on this environment, but with allies. But the NATO and EU alliances are in need of maintenance and extra efforts are required to guarantee their cohesion and sustainability. The signals sent from Washington by the current and former American presidents leave no doubt that most of NATO's European partners, including the Netherlands, must do more to honour the agreements made on the financing of their own armed forces. The United Kingdom's pending exit from the EU will complicate the military cooperation in Europe, when the EU's common defence policy was already lagging behind. All too often the EU's approach is based on crisis management.[1] The same ad-hoc approach is currently being used to protect the borders in the context of migration, albeit not in the classical territorial sense; that remains the core task of NATO. This crisis management distracts attention from what is really needed, which is consistent policy aimed at eliminating the underlying causes of the threats. The new types of threats to cyber security and infrastructure can also not be addressed in an ad-hoc manner.

The world order is shifting, the cooperation within NATO and the EU is under pressure and the Dutch armed forces are gasping for breath. What is the task of Dutch security and defence policy? There have been growing calls in the last few years to end the devastating cuts in the defence budget and to invest more. The acute threats and the conflicts in which the Netherlands is involved have served as a wake-up call. The shooting down of Flight MH17 over Ukraine, the streams of refugees from Syria and other countries, the conflict with Da'esh[2] in Syria and Iraq and terrorist threats make it clear that events in many of the world's flashpoints have a direct or indirect impact on the Netherlands. Conflicts in other countries have a spill-over effect in this country, as illustrated by tensions between population groups and the clashes over the Gülen schools after the failed putsch in Turkey on 15 July 2016 and over the constitutional referendum in that country. But how to ensure that any additional funds are not divided among the branches of the armed forces without any sense of strategic direction? What should a future-proof security policy that plots the course of defence policy look like? What strategic analyses should lie behind the political choices that are made?

1.2 A Different Security Policy in Turbulent Times

In this book the WRR takes the line that security policy must be based on strategic analysis. Defence policy and the role of the armed forces can then be determined on the grounds of that analysis. The discussion in this book expands on changes that have recently been initiated and on earlier policy recommendations. The policy letter *Turbulente tijden in een instabiele omgeving* [Turbulent Times in Unstable Surroundings] (2014) already underlined the need to anticipate a lengthy period of tension, both close to home and far away. That view was endorsed in the Ministry of Defence's multi-year perspective on the future of the armed forces entitled *Houvast in een onzekere wereld* [Certainty in an Uncertain World][3]: "Crises are succeeding one another in rapid succession and the diversity and complexity of threats and risks has also increased enormously. Internal and external security are more interrelated than ever. Europe is surrounded by conflicts and instability."

Other recent reports support that diagnosis. In a report entitled *Een kompas voor een wereld in beweging* [A compass for a changing world],[4] The Hague Centre for Strategic Studies (HCSS) identified four trends in the domain of international security of importance for the Netherlands: multipolarity, sovereignism, assertiveness and the 'Rise of the Rest'. In the Strategic Monitor 2017,[5] the Clingendael Institute referred to a world without historical precedent and described the current period as a 'multi-order' era. Significantly, the report of the authoritative Munich Security Conference 2017 was entitled *Post-Truth, Post-West, Post-Order?,* although it did end with a question mark. Like the other reports mentioned above, it found that the security environment is highly volatile but, above all, it observed that illiberal movements are gaining ground. The rise of so-called 'illiberal democracies' (as Fareed Zakaria called them in an article in *Foreign Affairs* in 1997), democratically-elected governments that stretch, and even go beyond, the limits of constitutional law by violating the fundamental rights and freedoms of citizens, form a threat to the international legal order that the Netherlands actively seeks to protect. More than ever, internal and external security, developments in the Netherlands and abroad, are interconnected.

These changes in the security environment and the threats to the international legal order not only call for a new strategic positioning, but also for consideration of increased investment in the armed forces. Accordingly, the government declared, in 2014, that it would adjust the level of ambition of the armed forces and, where necessary and possible, increase the financial resources allocated to the defence budget.[6] During the NATO summit in Wales in September 2014, the Netherlands alongside the other alliance members promised to raise defence spending closer to the NATO standard of 2% of gross domestic product (GDP) within 10 years. These undertakings were confirmed at the NATO summit in Warsaw on 8 and 9 July 2016, when the Dutch government indicated that it envisaged taking further steps on the basis of a long-term perspective, having regard to the international security environment and the budgetary scope.[7]

Although the need for additional budgetary room is repeatedly stressed, a problem is that up to now the commitments have been mainly statements of intent. To highlight the limits to the deployability of the armed forces, the Minister of Defence sent the NATO *Defence Planning Review* to the President of the House of Representatives of the States-General.[8] Without investments to increase their deployability, the Netherlands would be left with nothing more than a 'recuperating armed forces'. Although earlier austerity measures are being made up for, they will still not attain the necessary capacity.[9] In short, the proposals for additional investment are still a long way from meeting the requirements imposed by the security environment and the necessary strategic reorientation.

1.3 Security Policy Must Reflect Values and Meet Conditions

Since the 1899 Hague Peace Conference, efforts to secure peace through multilateral treaties and judicial settlement of disputes between states have been under way. The First World War crushed the expectations of those years, but after this horrendous war and again after the global disaster of the Second World War, responsible political leaders built and rebuilt the multilateral international legal order. The Netherlands has always been an active partner in these endeavours.

In our times, the notion of peace through international law is seriously challenged in at least two respects: on the one hand the re-emergence of nationalistic antagonism in authoritarian political systems, and on the other hand the growing importance of security issues that are insufficiently covered by international legal arrangements and dispute settlement: environmental issues, resource conflicts, human security and migration. The Dutch constitution requires the government to promote actively the international legal order. Given the requirements that result from a changing security environment, security policy and defence policy have to contribute to this mission in the national and international interest of peace and justice.

The tasks of the armed forces and defence policy are set out in general terms in the Constitution. However, in light of the changed security environment the ability to protect the integrity of national and allied territory, including the Caribbean parts of the Kingdom (the first main task) will have to be reviewed, in addition to the task of protecting and promoting the international legal order (the second main task). Assisting civil authorities in upholding the law, providing disaster relief and humanitarian relief, both nationally and internationally (the third main task) presents new challenges with the available capacity.

The capacity to provide security also has to be in balance with the values of a free and democratic society. Studies devoted to the issue of security warn of 'securitisation'.[10] Just as excessive 'medicalisation' transforms miscellaneous everyday complaints into medical conditions and 'juridification' leads to issues consistently being seen in legal terms and 'criminalisation' to issues being seen in the context of criminal law, 'securitisation' is a process by which various issues are clustered under the title of security, even when to do so is not always necessary or desirable. To put it another way, in a democratic society security policy must be embedded in a broader framework and care must be taken to avoid immediately treating every problem as a 'security problem' of a similar order.

The concept of security has already expanded in the Netherlands in the last decade. The Dutch Safety Board [*Onderzoeksraad voor Veiligheid*] was established in 2005, and the name of the Ministry of Justice was expanded to include 'Security' (currently 'the Ministry of Justice and Security'), even though security in the ministry's policy domain is primarily the result of effective law enforcement and therefore does not need to be named separately. This is a reflection of the attention to security in other ministries which are responsible for subjects such as food safety, the safety of the infrastructure, product safety and safety in public health. Caution is required, however, when security threatens to be seen as all-embracing.

The concept of security has itself gradually assumed a different meaning. The dynamics of technology, globalisation, geopolitics and the emancipation of society and individuals in relation to the state have changed the interpretation of the term 'security'. The scope of the term has been 'extended' from the traditional military defence of the state and its inhabitants against external threats, to the economic and ecological security of supranational regions and even the entire world. The term is also increasingly used in connection with the safety and future risks and threats to and within societies, social groups and individuals. Anyone wishing to explain and appreciate international views and policies on security has to be aware of the historical, cultural and political givens and future orientations that are inherent to this concept of security.[11]

1.4 Protection Cannot Be Achieved Through Isolation

This book is guided by the insight that the security of the Netherlands is connected to and entwined with that of the countries around us. It is pointless to think of this country's security in terms of entrenchment: purposeful policy can only be pursued by collaborating in measures to strengthen security in Europe and reducing the causes of dissatisfaction in Africa and the Near East. This insight is all the more relevant now that Brexit will create a gap in the development of the European security and defence policy and the likelihood that American involvement in NATO will diminish under President Trump – something that President Obama had in fact already clearly hinted at, particularly with a call for greater investment in the armed forces by the EU member states. Due to the *unpredictability* of the role of the US, the already escalating tensions with the Russian Federation assume even greater significance.

This book discusses at length the growing interconnectedness of national and international policy (including security policy) and the socialisation of security and defence and their consequences for international security and defence policy. These issues are discussed on the basis of an analysis of *international* security. For the Netherlands and Europe in particular, it is important to continue investing in strengthening the international legal order and in creating the conditions and circumstances under which countries on the eastern and southern flank of the EU can continue to develop in economic and political terms. At the same time, the WRR observes that the further development of the international legal order must be built on robust, realistic foundations in which the security of Dutch and European citizens is paramount. This means that the Netherlands must have a coherent international security policy, which includes future-proof armed forces.

The growing economic and political influence of countries like Russia, China, Indonesia, India, Brazil and Mexico is expected to translate into greater political tensions over trade, rivalry over access to raw materials and transport routes and expansion of potential military strength. The strain on the security of transport routes, the strategic game surrounding logistics and connections, as well as the status of networks and flow security, will increase. The international architecture under UN auspices will also come under further pressure. This means that the contours are emerging of a lengthy period of conflict for the control of, access to, availability of and safeguarding of vital interests, i.e., territorial, physical, economic and ecological security, political and social stability and the international legal order.[12] Consequently, complex conflicts and persistent global and regional problems such as climate change, migration, poverty and guaranteeing fair access to shared natural resources are difficult to tackle. The US has not lost any of its military dominance, but, due in part to its ageing population, internal political developments and pressure on financial and other resources, it is no longer willing and able to present itself as the sole hegemonic guardian of the international order. It is uncertain whether other stable alternative (regional) forums between groups of countries will emerge in time.

These power shifts and the fragmentation of power among state and non-state actors will probably also be reflected in the proliferation of nuclear weapons and other weapons of mass destruction in the coming years. Most nuclear powers are already modernising their arsenals. There are currently 25 countries in possession of nuclear material that can be used to produce nuclear weapons. The risks of further proliferation to states in the Middle East, of misunderstandings and accidents and of weapons falling into the hands of terrorists or criminals are therefore far greater than at the time of the Cold War, when five countries possessed nuclear weapons.

It is also becoming increasingly clear that the instability and potential for conflict on the southern and eastern flanks of the EU will persist in the coming years. Since the popular uprisings in the Middle East and North Africa (MENA) region in 2011 descended into large-scale destabilisation and refugee crises, Europe and the Netherlands have been experiencing the negative spill-over effects. There has been a sharp rise in the number of refugees and irregular migrants, which has caused a severe escalation in social and political tensions within and between the EU member states. The instability on the EU's southern periphery appears set to continue for the time being.

The states in the east of the EU and their immediate neighbours also appear susceptible to future instability and conflict because of their weak institutions, imbalanced economic development, corrupt political elites, ethnically diverse populations and difficult regime changes. They will possibly also be more exposed to Russia's disrupting influence. The armed conflict in Ukraine is a warning sign. Europe will have to be prepared for a scenario in which Russia tries to disrupt the Baltic States in the same way as it disrupted Ukraine and seeks closer ties with Serbia in order to increase its influence in the Balkans.

The multiplicity of (purported) risks and security threats, the interconnection of internal and external security issues and the greater involvement of social actors in security affairs all call for a clear policy. The government will have to explain the policy to the public. What are its priorities on the basis of the evaluation of the analysis of the internal and external environment in light of the national interests and values that are inseparable from the European context? What are the appropriate instruments and resources for the coming years?

Explaining them in outline is an essential ingredient of the public and political debate that is needed to create support for the necessary long-term investments in foreign and security policy. After all, the absence of sufficient support for policies increases the risk of vacillating policy and leaves less room for contradiction and the possibility for learning and innovation. Moreover, public and political debate about the goals, resources and instruments of security policy can provide a healthy counterweight to undesirable securitisation and permanent 'overkill' in the demands made on the government in the field of national and international security policy.

Most political parties in the Netherlands still regard the Netherlands' international orientation and its support for the international legal order and the rule of law as imperative: without them the country cannot undertake any action against threats such as international crime, weapons of mass destruction, uncontrolled migration streams or cyber crime. At the same time, however, there is also a reluctance to translate that realisation into a larger narrative, with clear review frameworks, envisaged goals and a coherent and consistent approach to problems derived from it.

However, it is the WRR's view that these narratives are essential for creating public support for the envisaged policy, particularly in times of international and European turbulence and public controversy. No one is immune to the images of streams of refugees and the heated local debates about providing shelter for refugees and asylum seekers. Many people in the Netherlands and Europe are more uncertain than they used to be about their prospects of a meaningful future. Immigration, globalisation and economic instability are strengthening the calls for individuality, small-scale interaction and the human dimension. Responding to these calls with guarded 'policy language' or financial discourse will not get one very far. Those responsible for defence policy and international security policy will have to be able to explain the strategic considerations in terms the layman can clearly understand, particularly when it concerns the importance of Europe, NATO and the UN and crisis-management operations abroad. Geopolitics and micropolitics go hand in hand. The mixed experiences of the last decade call for a realistic narrative, which has to encompass both awareness of the Netherlands' direct interests and the dilemmas it faces and the values that Dutch people wish to defend in the twenty-first century. Those experiences also suggest that human rights, human security and the social contract between state and society represent a challenge not only for distant developing countries. They must also be the subject of permanent maintenance and debate here in the Netherlands.

1.5 Approach and Demarcation

The aim of this book is to stimulate political and public debate about the priorities and capacities required for an intelligent, future-proof Dutch security and defence policy. The book is the result of a detailed study of the literature, interviews with numerous experts in the Netherlands and abroad, seminars, written input and peer review, in addition to the regular and intensive deliberations of the WRR members and the academic staff of the WRR. The book builds on previous reports by the WRR that are relevant to the subject.[13] The book does not contain a comprehensive analysis of trends or an elucidation of all the – theoretical - options for a new international security policy for the Netherlands, but it does consider the changes that the WRR feels provide compelling reasons for policy changes in the Netherlands. Radical policy changes, such as a return to the policy of neutrality preferred by Netherlands in the pre-1940 period or withdrawal from the EU, are not considered here. In the WRR's view, such breaks with the past are incompatible with the Netherlands' existing interconnectedness with the rest of the world and the country's international interests, values and orientations.

The WRR's point of departure is the necessity (including a constitutional obligation) and desirability of the Netherlands making an active contribution to the further development of the international system of security, liberty and legal order. The reference point for the action perspectives presented in this book is the armed forces. Many other parties are relevant to a broad security policy, but the question addressed in this book requires that it should focus directly on the role of the armed forces and defence policy. And because it concerns *Dutch* security policy, the book concentrates on the policy of the Kingdom of the Netherlands, which according to the Charter for the Kingdom of the Netherlands embraces the international relations and the defence of the entire Kingdom, including the Caribbean parts of the Kingdom. For the purpose of this study, *defence policy* is defined as the policy by which the government manages the organisation, instruments and finances of the armed forces. Defence policy is an essential component of the broader Dutch security policy.[14]

1.6 Structure of the Book

The structure of the book is as follows. The first part (Chaps. 2, 3, and 4) analyses the changing nature of security, the concept of security and the type of strategy formulation that is required. The second part (Chaps. 5, 6, and 7) contains a discussion of the consequences of that for defence policy and the position of the armed forces, and ends with the conclusions and recommendations.

In Chap. 2, the WRR analyses the extension of the concept of security that has occurred in the last few decades. The classical view of security as relating to national states and the protection of their territory still forms the core of thinking about national and international security. But security nowadays encompasses more than the protection of a state's territory against military aggression by another state. International security, for example, is increasingly linked to the security of society and the security of individuals (human security).

Globalisation, growing interdependence and geopolitical rivalry are prompting closer attention to economic security and flow security – safe, unimpeded flows of goods, services and data and the infrastructure required for them.

Chapter 3 shows that a steady expansion and socialisation of the security agenda is also occurring in the Netherlands. Attention in international security policy shifted from the Cold War dynamic to crisis management in fragile states, human rights and development. The Ministry of Defence and the armed forces translate this orientation into a sharp focus on 'expeditionarity', with missions carried out a long way from the country's own territory being linked to the agenda for development cooperation. The experiences with these missions have not been entirely positive, however. Economic security and flow security have only entered the vocabulary of the Dutch security agenda fairly recently and are not yet embedded as fully-fledged perspectives in the broader policy. On the other hand, the Netherlands does have a dedicated strategy for cyber security.

Chapter 4 outlines the rise of strategy formulation and national security strategies. Internationally, strategy formulation is a tried and trusted instrument for finding one's way in a complex, dynamic and unpredictable security context. By extension, it also leads to new policy processes and instruments, with appropriate changes in coordinating structures and forms of partnership. The Netherlands also studies the internal and external security environment, but has not yet fully embraced the available policy instruments and the underlying ideas.

Chapter 5 investigates the main tasks of defence policy in a changing security environment, starting with a brief sketch of the basic principles and recent priorities of Dutch foreign and security policy. There is then a discussion of the changing international security environment in the Netherlands and Europe, as well as the special requirements of the defence of the Caribbean parts of the Kingdom. The chapter describes how defence policy explicitly has to be seen in the context of the strong mutual dependencies between the national, regional and global levels. The main tasks are still the same, but have to performed differently, for example to meet the growing threat of terrorism as a result of Dutch interventions abroad and the persistent instability in Europe's surrounding regions.

Chapter 6 analyses the choices that have to be made to ensure the armed forces remain future-proof. The modernisation of the Dutch armed forces has been accompanied by significant spending cuts, rising operating and investment costs and wear and tear to equipment as a result of demanding missions abroad. There are deep-rooted problems as a result. The conclusion therefore has to be that the armed forces are not sufficiently future-proof. The problems have become even more urgent with the deterioration in the security situation around Europe, the agreements made in Wales (2014) and Warsaw (2016) to strengthen the alliance's defences and increase defence spending, and the increased vulnerability of the Netherlands and its inhabitants. This chapter describes the five conditions that a policy designed to create a sharper focus and make additional investments in the armed forces would have to meet.

Chapter 7, finally, presents the consequences of the analysis and presents the conclusions and recommendations for a future-proof Dutch international security and defence policy, and their implications for the armed forces.

Endnotes

1 Drent, M. & Zandee, D. (2016).

2 In this book the WRR uses the name Da'esh (Al Dawla al-Islamyia fil Irak wa'al Sham) rather than IS, the abbreviation commonly used in the Netherlands. Although the term IS ('Islamic State') has become commonplace, it reflects the incorrect claim of the movement's leaders and its supporters that such a thing as an Islamic State has been established by them, which is not the case. Furthermore, Islamic scholars have clearly distanced themselves from the term 'Islamic'. Although the WRR realises that the term Da'esh is less common, it has nevertheless chosen to use it for the reasons given here.

3 Ministry of Defence (2017: 5).

4 HCSS. (2017: 23–38).

5 Clingendael (2017: 2).

6 Parliamentary Documents II 2014–2015, 34,000, no. 1, 16 September 2014.

7 Letter to the House of Representatives with a report on the NATO summit in Warsaw on 8 and 9 July 2016, DVB/VD-156/2016.

8 Parliamentary Documents II 2015–2016, 28,676, no. 241, 24 March 2016.

9 Zandee, D. (2016).

10 Cf. Wæver 1995; Buzan, Wæver and de Wilde 1998; Balzacq, Guzzini, Williams, Wæver and H. Patomäki (2015).

11 Graaf, B. de (2012a).

12 Ducheine, P. (2016: 126).

13 Reports on, for example, Dutch foreign policy (Attached to the World [*Aan het*

buitenland gehecht], 2010b), Dutch development policy (Less pretension, more ambition [*Minder pretentie, meer ambitie*], 2010a), Dutch policy towards Europe (Rediscovering Europe in the Netherlands [*Europa in Nederland*], 2007), Dynamism in Islamic activism [*Dynamiek in islamitisch activism*], 2006) and economic resilience (Towards a Learning Economy [*Naar een lerende economie*], 2013).

14 See also the Dutch Defence Doctrine from 2013. Defence policy is part of the extended foreign and security policy.

Chapter 2
The Extension of the Concept of Security

Ernst Hirsch Ballin, Huub Dijstelbloem, and Peter de Goede

2.1 Security: A Tricky Concept

The classical view of security as relating to nation states and the protection of their territory still lies at the heart of thinking about national and international security. Defence in the sense of national defence is a key task of the nation state and for centuries the sovereignty of states has formed the basis of the international legal order. Today, however, security encompasses more than protecting the state's territory against military aggression by another state, as is illustrated in Fig. 2.1.

With globalisation and the emergence of new economic powers, for example, the relationship between economic and military security has become a far more prominent issue. In an interconnected world, flow security – safeguarding global flows of goods and services, infrastructural hubs and systems – is of the utmost importance. In a world of transnational connections, protecting national territory against hostile armies is not enough.

Moreover, since the 1980s, and to an even greater extent since the end of the Cold War, there has been a significant increase in concern for human rights and for economic and social development (*human security*). After all, physical violence is partially explained by the structural violence as a result of disadvantaged social circumstances. National and international security are therefore linked to the security of the society and the individual. The security of the nation state is not a goal in itself, but is for the benefit of the society.

Particularly since 11 September 2001, the blurring of the boundaries between internal and external security has also entered the debate about security policy[1] with the realisation that non-state actors such as Al Qaida and Da'esh have come to form part of the landscape of security and insecurity. The sovereignty of states over their

E. Hirsch Ballin (✉) · H. Dijstelbloem · P. de Goede
WRR, Scientific Council for Government Policy, The Hague, Zuid-Holland, The Netherlands
e-mail: ballin@tilburguniversity.edu; h.o.dijstelbloem@uva.nl; peter.goede@minbzk.nl

E. Hirsch Ballin et al. (eds.), *Security in an Interconnected World*,
Research for Policy, https://doi.org/10.1007/978-3-030-37606-2_2

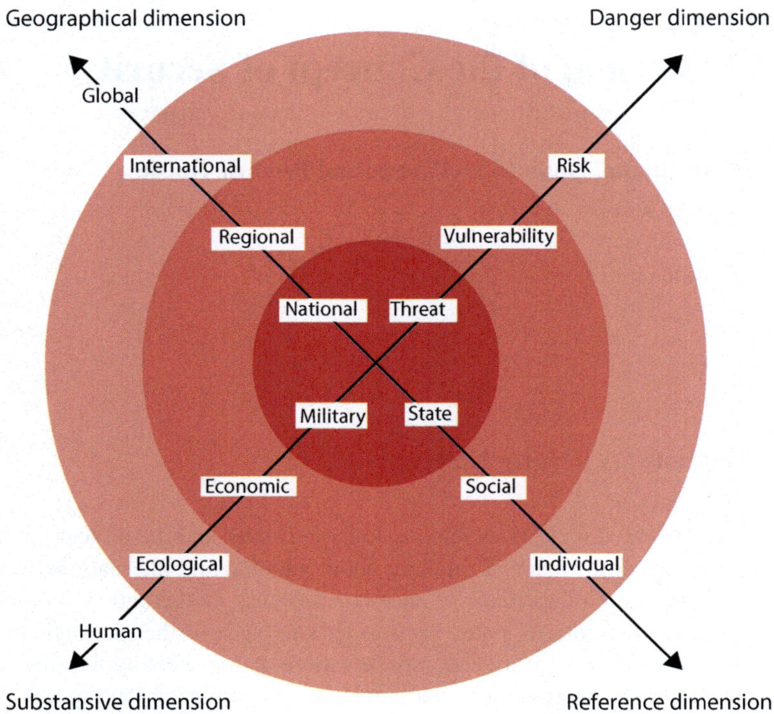

Fig. 2.1 Dimensions of security. (Source: DAASE 2013: 13)

own territory only offers limited protection against transnational networks of this type. An adequate geographic concept of security encompasses both the territories defined by individual states and the networks that transcend the borders between states. It therefore is possible to place international relations in a spatial perspective, provided that, in addition to national territories (protected by the internationally recognised right of self-determination), this perspective embraces transnational connections in which non-state actors also play a major role. Transnational issues such as migration, terrorism and climate change also mean that security is increasingly a matter for regional and international communities of states, such as the EU and NATO. Dutch security policy is part of European and North Atlantic security policy, although that does not mean that the Netherlands cannot set its own priorities within the security strategies of these alliances.

How the notion of threat is interpreted has also expanded. Security policy is no longer geared solely to the specific, known threat of armed aggression by another state. Nowadays, it also encompasses anticipation of diffuse, potential threats. The rise of risk governance[2] is part of a trend towards 'securitisation', by which we mean the enormous increase in recent years in concern for security, both in the sense of a desired situation and in the sense of policy or other activities aimed at achieving it,[3] as already mentioned in the introduction. Diffuse vulnerabilities and risks – and the associated emotions and feelings of insecurity – can in fact give rise to conduct that in itself becomes a factor in creating insecurity.

In other words, the meaning of national and international security has changed radically. In 2017, security refers to multiple substantive domains, reference objects, levels of geographic scale and interpretations of danger. Every dimension of the concept of security has expanded 'beyond' the classical view of security as it relates to nation states and the military protection of their territory. But that expansion is not a linear process proceeding step-by-step in a fixed sequence, but also has a non-linear dynamic. For example, the bipolar Cold War was followed by a unipolar world order, in which issues such as human security and the importance of multilateral institutions gained wider recognition, at least in many Western countries. But with the relative decline in the power of the West and the emergence of the BRICS countries – accelerated by recent developments in the belt of instability around Europe and the election of Donald Trump as president of the United States – a multipolar order now seems to be emerging in which the (collective) military defence of territorial integrity is once more growing in importance. Traditional power politics or geopolitics seem to have returned (if indeed they ever went away).[4]

In the following Sects. (2.2, 2.3, 2.4, and 2.5) we explain the concept of extended security against the background of historical developments in the second half of the twentieth century until the present day. The extension of the concept of security presents significant challenges for the government's security policy (Sect. 2.6).

2.2 The Substantive Dimension

The substantive dimension of security relates to the question: in which specific domains are security threats actually being observed? It concerns the type of security that security policy has to guarantee. Whereas military threats dominated the security discourse in the 1950s and 1960s, economic, ecological and humanitarian issues have come to the forefront in the succeeding decades.

In addition to the 'realistic' view, in which military threats to the state's security are central – and which is characteristic of the period after the Second World War and the Cold War – since the oil crises in the 1970s there has been a growing awareness of a new threat: in an economically integrated world, economic security is at least as important as military security. Later, with the publication of the Brundtland Report in 1987, ecological threats to security came to prominence. Nowadays, we speak mainly of the threat to 'planetary security' from the consequences of climate change. After the Cold War, the concept of security expanded further with the attention for 'human security' and human rights, although in the decade since 2008 (the year of Russia's war with Georgia against the background of the prospect of NATO membership for Georgia) traditional considerations of power politics have again come to the forefront.

2.2.1 Economic Security

In the 1950s and 1960s, international security policy was primarily concerned with protecting the security and sovereignty of the nation state against external military threats. Unlike military security, the term economic security was not yet in common usage. In the literature on international relations, economic security was usually narrowly defined as a country's capacity to ensure it has sufficient economic resources to sustain its own security, military or otherwise.[5]

The interest in the economic security of states received a substantial boost with the oil crises in the 1970s and the pursuit of strategic trade policy by emerging powers from the 1980s. These trends made it clear that imports of energy and strategic commodities and the general stability of the global trading system represent crucial interests.[6] Some experts argued that military, economic and political instruments should be incorporated into a more integrated approach to security, including a strategic policy on raw materials. In the US, the two oil crises led to a securitisation and militarisation of the country's policy towards raw materials. The Carter Doctrine (1980) made it clear that the US would use military means to safeguard its own national and international economic interests, not only in the Middle East but also elsewhere.[7] Globalisation and the rise of new economic powers gave a further impulse to the integration of economic and military security. The interconnectedness of global production chains means that severe disruptions of those chains occur more frequently, with the accompanying risks of cascade effects. A well-known example is the disruption of the global supply chain that occurred after the earthquake and tsunami in Japan in 2011. The event badly affected production at Toyota and Honda, but major economic 'aftershocks' were also felt elsewhere in Asia and in Europe and the US. Similar cascade effects with interruption of industrial production occurred after the eruption of the volcano in Iceland in 2009.[8] The vulnerability of global production chains lies mainly in the fact that the production of key components is concentrated in a small number of locations.[9,10] Emerging economies with only modest financial buffers are particularly vulnerable to this type of disruption, particularly if they are dependent on just a few sectors or activities.

Figure 2.2 gives an indication of the importance of economic security on the basis of the multilingual MetaFore Approach used by The Hague Centre for Strategic Studies (HCSS). The figure contains the number of times and the frequency with which different economic aspects of security appear in hundreds of reports on security in the different language regions. Because a large proportion of international trade in industrial products consists of semi-manufactures, countries have become more dependent on one another. The end product consists of many components sourced from numerous locations and sometimes a product is exported and imported again several times before reaching its final stage. A notable finding is the large number of references in non-Western reports. The availability of water is naturally an important subject in the Arab world, while the financial system plays an important role in the Western sources.

	Arabic	Chinese	Farsi	Russian	Turkish	Western
Economic	9 (20.5%)	30 (50.0%)	35 (83.3%)	15 (48.4%)	10 (58.8%)	4 (26.7%)
Global Logistics and Trade	1 (2.3%)	13 (21.7%)	2 (4.8%)	2 (6.5%)	4 (23.5%)	5 (33.3%)
Financial system	0	5 (8.3%)	3 (7.1%)	13 (41.9%)	2 (11.8%)	2 (13.3%)
Water	18 (40.9%)	1 (1.7%)				
Food	10 (22.7%)	4 (6.7%)				
Natural Resources		5 (8.3%)	2 (4.8%)	1 (3.2%)	1 (5.9%)	1 (6.7%)
Vital Elements	6 (13.6%)	1 (1.7%)				1 (6.7%)
Labour Market		1 (1.7%)				1 (6.7%)
Health						1 (6.7%)
Total	44 (100%)	60 (100%)	42 (100%)	31 (100%)	17 (100%)	15 (100%)

Fig. 2.2 References to 'security'. (Source: The Hague Centre for Strategic Studies, Strategic Monitor 2016 *The Wheel of Fortune,* 2016*: 160–161*)

Economic security has been gaining in prominence in the last few decades and a broad definition of the term encompasses security in the domains of cyberspace, maritime affairs, raw materials, energy, food and ecology. So-called 'non-traditional', more comprehensive security studies focus predominantly on these themes.

The challenges connected with the increased interconnectedness and transnational spill-over effects are also sometimes clustered under the title of flow security, a term that was coined in 1972 by the economist Michael Adelman in the context of the vulnerability of the American oil supply during the first oil crisis.[11] The term flow security expresses the reverse of territorial security. Although initially the concept did not really catch on, about 10 years later it was also being used in relation to the electricity network and electronic data traffic. Flow security is essentially concerned with the risk of supply being threatened by an interruption at a certain point in the chain. In the case of consumer goods, investments, money, (digital) information, tourists and employees, it relates to functional systems that have consciously been created to meet specific needs of society. However, there are also systems with flows that are undesirable or have negative consequences, such as irregular migration, drug trafficking and crime.

Flow security encompasses the capacity to prevent undesirable flows (or interruptions to essential flows) or to deal with them in an acceptable manner by mitigating their negative effects. There are two major challenges in addressing flow security. First, investments in the infrastructure must be sustained in order to prevent blockages, leaks, loss of quality, contamination, etc. in the channels and hubs used by the flows. Second, and by extension to the first, flow security requires that all the actors involved in the system possess the capacity to anticipate and adapt – for example, the ability to deal with internal system failures that interrupt supply or with damage caused by actors or influences external to the system. Another example is the management of a global public good like cyberspace. Cyber security is high on the agenda of a growing number of countries and companies because of the growing frequency with which they are subjected to digital attacks, espionage and other forms of cyber crime.[12] Furthermore, they increasingly regard the Internet itself, its infrastructure and central protocols, as a legitimate instrument for promoting their own strategic interests. This is usually at the expense of the public core of the Internet.[13] The security of the physical infrastructure on the seabed that keeps this virtual world intact, is equally crucial. Approximately 95% of all intercontinental communication depends on the 600,000 miles of submarine fibre optic cable and some 24 cable landing points. An attack on this infrastructure could cause enormous damage for the global economy because of the total reliance of worldwide electronic data traffic on it.[14]

2.2.2 *Ecological Security*

The Brundtland Report already drew attention to the security aspects of environmental problems in 1987. More recently, climate change has pushed the ecological threats to security higher up the political agenda. With the focus on the potential for climate change to cause conflicts, the concept of security has expanded once again.

The economic growth of countries in Asia, the Pacific region, Latin America and Africa, the worldwide population growth and the rise of a global urban middle class are driving an enormous increase in demand for energy, water, food, minerals, land and other natural resources.[15] Safeguarding a steady supply of energy, being able to cope with price fluctuations and reducing vulnerabilities by diversifying and making the transition to renewable energy sources are urgent challenges, especially in light of the persistent crises in the Middle East and North Africa and the tensions with Russia.[16] The Atlantic Basin is likely to become increasingly important because roughly 60% of the increase in oil production until 2030 will come from there. Estimates suggest that the international trade in energy will have doubled by around 2050 and increasingly involve liquid gas rather than oil. Furthermore, raw materials extraction and the trade routes in the Arctic region will grow in importance because of climate change. Some even argue that the Bering Sea will become a formidable competitor of the Panama Canal for the transport of energy.[17]

These trends will also further increase the migration potential in the world, with more of the people concerned coming from unstable and weak states confronted with protracted internal and regional conflicts, as is now the case in Syria, Iraq, Afghanistan and Libya, or with combinations of conflict, drought and scarcity of raw materials and food, as in Yemen and some Sahel countries. The OECD forecasts that by around 2030 almost half of the world's population will be confronted with the negative effects of the rising sea level and that this will cause more people to seek their fortune elsewhere.[18] Unless mankind takes effective action against climate change, the UN High Commissioner for Refugees has estimated that between 250 million and 1 billion people will be forced to leave their own countries over the next 50 years. The poorest and most vulnerable groups in the developing countries of Africa and South and East Asia, and some small island states, will be hardest hit. This explains why climate security is high on the research and policy agendas of many governments, companies and international organisations.[19] Although recent developments in biotechnology offer possibilities for improving food production, breakthroughs in gene technology could also pose a threat because they facilitate new forms of warfare, for example by allowing the genetic composition of organisms to be quickly altered using so-called gene-drive technology to give them properties that are harmful to humans (*Trendanalyse biotechnologie 2016* [Analysis of Trends in Biotechnology 2016]).

2.2.3 Human Security

After the end of the East-West conflict, attention shifted to conflicts within societies and to the security of social groups and individuals. The concern for protecting people in safe zones in crisis situations, and more generally protecting basic human rights, once again created a new domain in the security discourse: human security. Nevertheless, the thinking in terms of humanitarian security has been under pressure in recent years due to the relative waning of the power of the West. There seems to be greater reluctance to carry out stabilisation operations with a humanitarian goal.

Central to the concept of human security is the protection, freedom and development of individuals in society[20] and therefore closely related to a full realization of human rights. In the 1990s, humanitarian disasters in Somalia, Rwanda and Bosnia and the civil war in Kosovo fuelled the debate about whether the international community, and the UN in particular, had done enough to prevent those serious crimes against international law and human tragedies or at least end them sooner.

The concept of human security embodies the post-1990 thinking about development issues. It initially comprised two different, partially competing schools of thought.[21] The first advocated the ending of the arms race and the use of the money spent on defence for development purposes. After all, poverty and economic hardship form a breeding ground for violence in a number of regions. Structural development aid could ultimately remove this root cause of violence. This appeal was later fleshed out in the *Human Development Report 1994*, which described human security as "… concerned with how people live and breathe in society, how freely they exercise their many choices, how much access they have to markets and social opportunities – and whether they live in conflict or peace".[22,23] Such a fairly comprehensive interpretation of human security – which was adopted by the Japanese government, among others – has been described as an approach aimed at creating a society that is free from fear and provides adequate social security.[24] In this approach, attention is devoted both to addressing chronic problems like hunger, disease and oppression and to providing protection in conflict situations and crises.

The second school of thought – which is propagated by the Canadian government, among others – narrowed the concept of human security, partly for pragmatic reasons, to the protection of individuals against direct physical violence. In practice, this translated into diplomatic efforts and humanitarian interventions to tackle violent conflicts and violations of human rights.[25] The so-called Brahimi Report (Report of the Panel on United Nations Peace Operations 2000) called for greater coherence in the actions of various UN organisations during peacekeeping operations and introduced the concept of 'protection of civilians'. In the report of the International Commission on Intervention and State Sovereignty in 2001 and in a later amendment passed during the United Nations World Summit in 2005, this reasoning was fleshed out into the concept of Responsibility to Protect (R2P), which enunciates the principle that every actor – whether it is a state or an international organisation like the UN – is obliged to protect the civilian population, by military means if necessary, against 'genocide, war crimes, ethnic cleansing and crimes against humanity' if the state where the atrocities are taking place fails to do so.[26,27] This interventionist element of human security was severely criticised by countries in Asia and Latin America, as well as by the US and Russia. They saw it as an infringement of the principle of non-intervention. The southern countries, in particular, regard it as legitimising military interventions by strong states in weaker states on the basis of their national interests.[28] Furthermore, R2P goes further than the original *right* of humanitarian intervention by referring to the *responsibility* to intervene. According to these countries, this would significantly lower the threshold for military interventions (Box 2.1).[29,30]

Box 2.1: Human Security in the UN and the EU

In formulating the Millennium Objectives in 2000, the UN tried to codify human security and make it quantifiable. In 2012, the UN General Assembly adopted a resolution based on a broad definition of human security as 'freedom from fear, freedom from want and freedom from indignity'. Human security is an approach that can help the member states to address the broad challenges to 'the survival, livelihood and dignity of their people'.[31] The 'human security' approach also resonates in the Sustainable Development Goals (SDGs) that the UN adopted in 2015.

At EU level, human security is closely connected with the efforts by the former High Representative for Foreign Affairs and Security Policy Javier Solana to give the EU – with the war in Iraq fresh in the mind – a stronger, more international and above all multilateral role in security that differed explicitly from America's unilateralism. The European Security Strategy (2003) did not mention human security, but did expressly state that military responses alone are no longer sufficient in light of the new challenges facing the international community and that there was a need for a comprehensive approach with a balance between civil and military efforts. The so-called Barcelona Report in 2004, A human security doctrine for Europe, elaborated on this and proposed the creation of a 15,000-strong human security response force.[32] This was followed by the report entitled A European way of security (2007), in which the human-rights dimension of human security assumed an important position.[33]

The Global Strategy on the EU's Foreign and Security Policy (June 2016) was drafted in a drastically altered geopolitical and security context. In it, the EU chose for 'principled pragmatism' as the guiding principle of the EU's external action. Human security, and an integrated approach to conflicts and crises derived from it, is one of the five priority areas. In this context, 'integrated' refers to the use of all the policy instruments at the EU's disposal, but also to action in every phase of the cycle of conflict and at various levels of governance and with various partners.[34]

The adoption of R2P at the UN's World Summit in 2005 was welcomed by its proponents as a victory for human security and a humanisation of the international legal order.[35] But there was also criticism. The first objection was that the securitisation of development and human rights can all too easily end in military interventions and 'just war',[36] which could actually reinforce the insecurity (or feeling of insecurity) of citizens and bring the UN's human-rights regime into discredit in regions and among population groups that are on the 'receiving end' of such actions. One example is the American 'war on terror' in Afghanistan and Iraq in the wake of the attacks on 11 September 2001, which was accompanied by grave violations of human rights and created new breeding grounds for regional instability and insecurity.

That leads on to the second fundamental criticism, namely that up to now human security has been used mainly in the context of North-South relations, for interventions and stabilisation and crisis operations in weaker or fragile states. Consequently, the human-security discourse is said to primarily strengthen the existing, Western-dominated power structures in the international order. The claim that human security is a universal challenge then loses cogency.[37]

The different approaches to human security have meanwhile converged somewhat in the policy discourse of the UN, the World Bank, the EU, NATO and the Organisation for Security and Cooperation in Europe (OSCE), but also of many nation states including the Netherlands.[38] This has two major consequences. The first is that human security has greater potential than formerly to function as a corrective of a narrow state perspective when the classical social contract between state and citizens is under pressure and the protection of human rights demands urgent attention.[39] The second consequence is that human security is nowadays given a broader and more inclusive interpretation within these international organisations. As a result, the needs and wants of ordinary citizens can also be spotlighted. There is also greater attention to the underlying causes, and hence for a preventive, integrated approach to tackling violence, insecurity and instability. In short, human security has drawn development and humanitarian issues into the traditional domain of state security, but also brought security into the domain of non-state actors in the field of development and humanitarian aid.[40] Accordingly, the United Nations Organization has accepted the view that a human security approach is indispensable to accelerate progress on the Sustainable Development Goals.[41] The 2030 Agenda for Sustainable Development lists Sustainable Development Goals. The 16th of these goals in particular – on 'peace, justice and strong institutions' – is plainly related to the overreaching aim of international peace.

2.3 The Reference Dimension

The reference dimension of the concept of security turns on the question of whose security must be guaranteed: the nation state's, the society's or the individual's. Here too we see an extension of the concept of security beyond the state as the referent object of security policy.

2.3.1 From State Security to Social and Individual Security

The realistic school of thought on security is state-centric: the priority is the security of the nation state. The security interests of other referent objects (specific groups or individuals) are deemed to coincide with that. Critics of the realistic tradition in international relations have pointed out that it is in fact generally states that threaten the security of non-state actors and individuals. Particularly after the end of the Cold War, a view of security emerged that gave more weight to social development and individual human rights.[42] The concept of security then centres not only on physical and inter-state violence, but touches on other types of threats to communities and individuals.

At the height of the Cold War, there was a large degree of consensus between the state and the people in the West. In return for protection against the (nuclear) threat from the Eastern Bloc and protection of their personal liberties, citizens gave their loyalty to the authority and the monopoly on the use of force of their nation state. When political leaders lose sight of this crucial dimension of the relationship, security policy becomes cynical realpolitik and the interests of the state come to dominate exclusively.[43] As we saw in the preceding paragraph on human security, a functioning state is necessary for the security of the population, but a (failing) state can also endanger the security and development potential of its own citizens. Accordingly, the referent object of security – at least in Western democracies – shifts from the state to the society and the individual. In that context, 'societal resilience' is a theme that is receiving growing attention.[44]

The extension of the referent object means that the traditional role of international law in promoting and guaranteeing peace and security also needs to be expanded. Terrorist crimes by non-state actors fall under international or national criminal law, while it is generally accepted that international humanitarian law also applies to national conflicts between combatants that cannot sign up to the Geneva Conventions. However, the broader sources of international security are not confined to restricting methods of warfare and prohibiting wars of aggression. Far more attention and resources need to be devoted to the human rights laid down in the Vienna Declaration of 1993 if they are to be genuinely relevant for removing and preventing (feelings of) deprivation and creating the prospect of a meaningful life. In the Near East a new large-scale military intervention would further exacerbate the feelings of frustration, especially if it was again followed by further chaos – as previously in Iraq and in Libya – after a political revolution. Longer-term stability calls for a deeper engagement with social and economic development. The Advisory Council on International Affairs (AIV)[45,46] referred to this in a advisory report on security and stability in Northern Africa: "The AIV encourages the government to tackle underlying causes. Africa's rapid population growth requires paying attention to the complete range of human rights described in the Vienna Declaration. Serious deficits in the observance of fundamental rights to education, healthcare and economic development severely exacerbate the problems described in this book, and it is essential to address these negative developments in the preventive approach that the AIV considers necessary in Northern Africa".[47] Hence the following recommendation by the AIV with regard to EU policy: "The EU member states should make promoting stability and security - and human security in particular - in Northern Africa one of the main aims of European foreign and security policy in the coming period, together with responsible economic development, political reform and respect for universal human rights".[48]

2.4 The Geographical Dimension

The geographical dimension of the concept of security determines the levels of geographic scale to which security policy applies. Is it the territory of a state, based on the idea that domestic and foreign policy are strictly separate and that the limits of the state coincide (or should coincide) with the borders of the society? This defensive 'container model' of security has been inadequate for a long time. The expansion of the geographical dimension has drawn more attention to regional, international and global security complexes, in which states are so interconnected that their national security cannot be considered separately from the security of other states.[49] Security is then no longer a hallmark of a national territory, but a relational, inter-state concept. Moreover, security and insecurity are manifested more than ever in networks of non-state actors which cannot be classified as 'internal' or 'external', as international terrorism consistently demonstrates. Regional, international and global interrelationships demand investment in collective security and supranational alliances.

Security has a particular geographic or territorial connotation. Article 8bis of the Statute of the International Criminal Court defines an act of aggression as the use of armed force by a State against the territorial integrity of another State. On a smaller scale, trespassing and disturbance of public order cause insecurity. For many centuries the state performed its task of guaranteeing the general safety of persons and goods by guarding and defending the territory of the state and by preventing and ending incidents of public violence and other breaches of the peace. Fortifications, watchtowers and watch keepers typically personified the state's security task; respect for national borders and the boundaries of a person's property were manifestations of peace abroad and at home.

By their nature, promoting security and combating insecurity still extend to a particular physical area used by people. Defence is essentially the defence of the territory of the Kingdom of the Netherlands and, as a member of NATO, that of the member states north of the Tropic of Cancer.[50] This accords with the idea that the sovereignty of states is in principle circumscribed by their physical land and maritime borders.[51] A properly functioning international legal order, as it has grown with ups and (deep) downs since the First Hague Peace Conference in 1899, guarantees international peace because states respect one another's borders and jurisdictions. It is one of the tasks of the UN Security Council and the International Criminal Court to ensure they do, a task they can only perform if states genuinely cooperate. The Dutch armed forces, however, are not there solely to defend and protect the interests of the Kingdom, but also to enforce and promote the international legal order.[52]

2.4.1 Security in a Regional, International and Global Context

In the annual report for 2002, the former Minister of Defence wrote: "The high level of deployment in the last year underlines the continuing need for well-trained and well-equipped armed forces. In light of current international developments, in which the distinction between internal and external security is becoming increasingly blurred, that need will not diminish in the future".[53] The minister was referring to military contributions to the peace and reconstruction process in various countries and to the war against international terrorism. Many others have also noted that internal and external security are interrelated, often with reference to the transnational impact of civil wars and terrorism. Security and insecurity manifest themselves more than they used to in networks that cannot be classified as 'internal' and 'external'. International terrorism testifies to that, as does the – still inadequately restrained – international arms trade. Unrestrained internal conflicts like those in Ukraine and Syria had an enormous international impact, in one case in the form of political and military interventions by a neighbouring country, in the other through the exodus of large numbers of refugees to other countries in the Near East and Europe. The international risks ensuing from Da'esh have also arisen from irreconcilable differences in Iraq and the Levant.

These developments have led some to rediscover 'geopolitical' models of security policy, in which the emphasis is entirely on a 'realistic' approach to external security, in contrast to models that link internal and external security. However, a closed, defensive approach to the relationship between states with 'sovereignty over their own territory' – defence as a bulwark against external forces – is actually at odds with economic and cultural integration, and increases the risk of conflicts when it evokes nationalism and xenophobia.

Franklin D. Roosevelt already established the relationship between internal and external security in his State of the Union address on 6 January 1941: the relationship between respect for fundamental rights – the four freedoms – within states and peaceful relations between states. This is why the promotion of the international legal order (in the Netherlands, a constitutional obligation: see Article 90 of the Constitution) is an essential element of an effective peace and security policy.

In the absence of peaceful and prosperous internal development, the chance of international conflicts increases. But efforts to create internal peace have generally had less of a preventive effect than expected. They were often a vain attempt to control a crisis - and were sometimes even counter-productive because the intervention was associated with a party to the conflict that was not seeking reconciliation, as in the case of the regime in Iraq which is dominated by a single population group. Rather than being intensified, genuine prevention by investing in the social and economic development of 'fragile states' declined in the wake of the need for austerity measures – particularly in the years after the financial and economic crisis in 2008. Making the situation worse was the fact that traditional military capabilities were also substantially reduced, which meant that exerting pressure on dubious regimes lacked credibility and the sense of insecurity on the peripheries of NATO and EU territory grew.

The enormous increase in the number of non-state actors[54] is closely connected with the end of the Cold War, globalisation and processes of democratisation since the 1990s. With the liberalisation of the flows of goods and services and financial flows and the lifting of stringent barriers to the movement of persons, companies, media and NGOs can move more easily around the world and have quicker and easier access to capital and manpower. Cheap information technology further means that even the very smallest organisations and movements can portray themselves as actors not bound by national sovereignty.[55]

What does this trend mean for the international context and security? Just as globalisation is not by definition favourable or unfavourable, the growth in the number of non-state actors is not inherently a good or a bad development in security terms, nor does it automatically undermine state authority. At the same time, it is clear that the combination of the diversity of actors and the ongoing technological developments makes it more difficult to determine where threats are coming from.[56] The cyber technology, biotechnology, robotics and drones that can make life more pleasant, can also fall into the hands of ill-intentioned non-state actors.[57] Increasingly, these groups organise themselves as illegal networks that are difficult to track and make smart use of global chains and flows.[58] This category includes terrorist groups such as Al Qaeda, Boko Haram and Da'esh, which undermine weak state institutions and the precarious security situation of citizens in fragile states (see Fig. 2.3). Furthermore, like states they use both 'hard' military means and 'softer' instru-

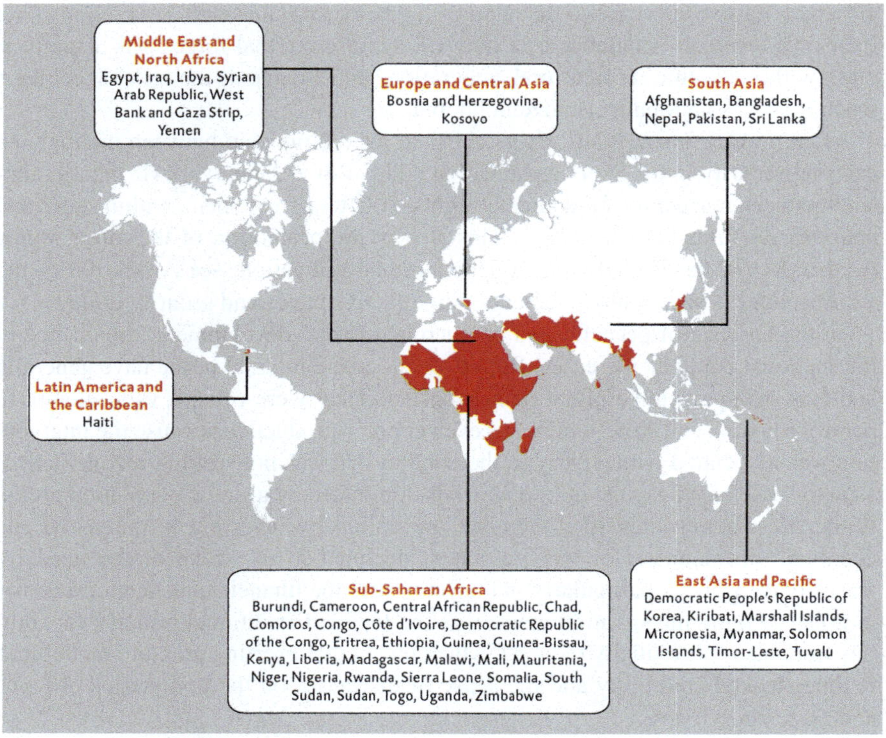

Fig. 2.3 Fragile states. (Source: OECD, *States of fragility 2015*: 15; other states, such as Kazakhstan, Uzbekistan and Venezuela, could also be added to the list)

ments, such as media, social media and the Internet, to bolster their legitimacy and to win the hearts and minds of potential supporters.[59]

Privatisation of security is a trend that can be seen in the many fragile states afflicted by conflict (see Fig. 2.3), but also in developed countries.[60] As a result, the boundaries between state and non-state security actors are blurring. The current reality is one of a sliding scale from more to fewer state services and actors in the field of security.[61,62] The private security industry has been growing since 11 September 2001 when (state) security rose higher on political agendas and a growing number of national armies, NGOs and companies started outsourcing security tasks. Private Military Contractors played an important role (from providing catering and guarding oil plants to organising transport and interrogating prisoners) in Afghanistan and Iraq, for example. Governments often hire private contractors for reasons of cost control and flexibility.[63,64]

There is also increasing evidence that future inter-state conflicts will be played out in the free, still open spaces of international waters, the ocean floors with their deposits of minerals and fossil fuels, fishing grounds and transnational waterways and drinking water reservoirs. The potential for conflict is particularly great where there are competing claims to access, control and ownership, but where there is no political willingness to embrace rules and dispute-resolution mechanisms for such global public goods or global commons. Closer to Europe, this mainly concerns the eastern part of the Mediterranean with its large gas reserves, the Arctic region with its oil, gas, minerals and fish, and the Horn of Africa and Sub-Saharan Africa. In these regions, the scarcity of water and energy and lengthy periods of drought due to climate change are taking their toll in the form of regional conflicts and increased migration.[65] At the same time, the traditional tasks of a state's security and defence policy, such as defending against violations of the integrity of the state's territory, air space and territorial waters, deterring the enemy and preventing nuclear proliferation in various regions of the world, including Europe, remain as necessary as ever. This all implies the need for a simultaneous strengthening of more defence-oriented and prevention-oriented elements of international security policy.

2.5 The Danger Dimension

The most recent expansion of the concept of security relates to interpretations of what constitutes danger. It depends on how specifically or diffusely danger is defined. It makes a great deal of difference whether security is viewed as the absence of a military threat to a clearly-defined territory or as the reduction of economic vulnerability in a globalised world. Or as the reduction of risks and insecurity even before there is any question of an acute threat. The conceptualisation of security as defence against a military threat was closely connected with the realistic theory and the Cold War era, when hostile states confronted each other in a clear balance of deterrence. However, security is increasingly also connected with the far more elusive violence of non-state actors. It is even linked to diffuse dangers without any identifiable actor with hostile intentions. Subjective perceptions – including feelings of hope, humiliation or fear – play an important role in the securitisation of threats, vulnerabilities and risks.

2.5.1 Securitisation of Threats, Vulnerabilities and Risks

By 'securitisation' is meant that political actors in nation states actively engage in the 'top down' framing of tensions as threats to security and securing institutional and public support for their views.[66] By establishing what the threats are, how they manifest themselves and who or what is threatened, they give meaning to the world around them and establish order in the international environment. However, securitisation also arises from society itself, particularly when there are fears involved. Societal actors can themselves draw attention to specific issues and urge politicians to take prompt action and adopt tougher security measures. Thanks to the proliferation of actors in civil society, the privatisation of some security tasks and the ever-present media, the question of who can place which threats on the security agenda is increasingly the outcome of a wider public debate. Politics and governance are then part of a wider agenda-setting process.[67] In other words, security is a social construct: it is the result of a process in which an actor is able to convince an audience that a referent object is threatened and has to be protected.[68]

This leads to two observations. First, a wide range of narratives concerning security can exist alongside one another in the society. The problems that are placed on the agenda and addressed, and how, can vary depending on the time and place. Some civil-society actors are simply better than others at getting across their interpretation of security or insecurity. Second, certain interpretations - including those of select individuals in the society – can become dominant and anchor themselves in institutions and the collective consciousness for lengthy periods. They find their way into more formal strategic documents, parliamentary papers and military doctrines, but also into folklore and popular wisdom.[69]

This means that what 'society' or particular communities and networks describe as urgent security issues are not necessarily more objective, more morally just or more rational than those identified by the 'state'.[70] For example, the security interests articulated by some population groups in multi-ethnic states could harm the security interests of other ethnic groups. And in some societies, militant nationalism, the demand for self-determination or calls for stricter government action against illegal migrants are in fact sources of actual or unrealistic insecurity.

What this also means is that the government cannot automatically be guided by what the public regard as urgent or relevant. After all, it is the government's task to represent the wider public interest and protect the freedom and equality of every citizen. That is a two-fold task. On the one hand, it calls for the creation and preservation of public support for sustainable long-term investment in the necessary (military) security capacity, even if it has no immediate or visible added value for people's day-to-day lives in the short term. On the other, it is necessary to guard against the government being permanently expected to meet excessive demands in the field of security.[71] After all, in many countries the attention to public safety leads in practice to a strong focus on anticipating exceptional events that could disrupt society. As a result, a variety of risks and uncertainties come to the fore, which can lead to over-reaction and imbalances in the security policy.[72] Figure 2.4 shows the deployment of the armed forces in the performance of a number of public security tasks other than the traditional task of protecting the territory and the public against threats from other countries.[73]

The question is what effects focusing on a wide range of risks, uncertainties and worst-case scenarios have on the perception of security and civil liberties. The socialisation of security can lead to increased pressure on the government and civil-society actors to take measures to address a perceived '*feeling of insecurity*'. Those feelings of insecurity can then be sustained if the measures are further magnified in the media and on social media. An expanded, inclusive security agenda could then even cause security to become a goal in itself rather than being a prerequisite for further social development which therefore has to be weighed against other political priorities.[74]

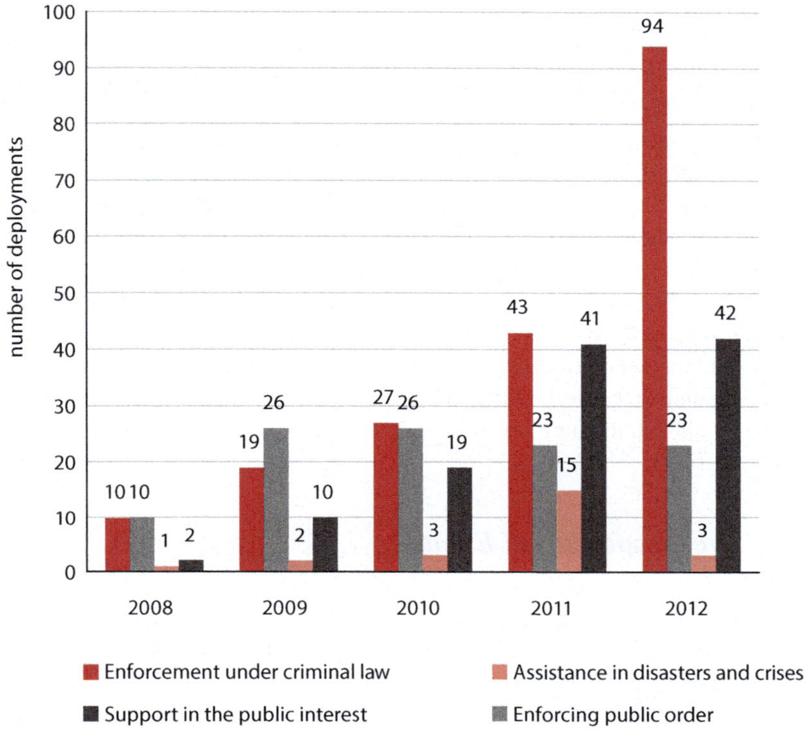

Fig. 2.4 Deployment of armed forces to provide ad-hoc assistance. (Source: Clingendael, Strategic Monitor 2014, A World Order Balancing on the Brink [*Een wankele wereldorde*], 2014: 105–107)

Violent conflicts prompt the most direct feeling of insecurity, which is why when one speaks of 'security policy' thoughts normally turn principally to the area of operations of the armed forces and of the police and other law-enforcement agencies. Since 1945, and for more than a century preceding the Second World War, the Netherlands has not been involved in military conflicts on its own territory on the European continent. So far – at the time this book was being written – the Netherlands has also been spared the large-scale terrorist violence that has recently afflicted a number of European cities. Nevertheless, there are many who feel that there is a greater threat of international and national insecurity at this point in time. A distinction is sometimes made between security as 'objective circumstances' and 'subjective feelings of security', which are easier to measure in surveys.[75] The former refers to the absence of actual threats to our way of life, the latter to the absence of a fear of that way of life being undermined. Furthermore, the threat can relate to attainments of a material nature (for example, access to work, housing or education), but also to idealistic attainments (e.g., values and customs that are taken for granted).

In the last few years the feelings of insecurity have been compounded, in the Netherlands and elsewhere, by violent incidents that do not bear the hallmarks of aggression by another state. Neither the shooting down of Flight MH17 nor the capture of areas in the Near East by Da'esh nor the attacks in Paris, Brussels and Berlin correspond with the traditional image of military conflicts between states seeking to control parts of each other's jurisdiction, in other words territory. Nor can these events be – literally - mapped in the former customary manner. International security (or insecurity) increasingly clearly displays the characteristics of a 'hyper-connected world'.[76] The view that the conflicts between regimes and rebel movements in Iraq and Syria were domestic conflicts that did not affect the states in Western Europe, for example, proved very short-sighted. Not only were these armed conflicts nourished by financial support and arms sales from other countries, some of the fighting itself has moved to cities in Europe with the terrorist activities of Da'esh. But impulses are also emerging from the European cities themselves to move and connect battlegrounds, particularly where radicalised individuals are recruited by Da'esh to come and fight in the Middle East and for terrorist activities.

2.5.2 The Geopolitics of Emotion

Another important perspective that can be added to the analysis of the entanglement of national and international security is thinking in terms of emotions. Personal humiliation, for example due to the structural absence of the prospect of securing a livelihood and of personal development, but also at collective level, such as Russia's loss of international stature in the 1990s, play an important role in this respect. Fears, particularly among groups that come under attack because of their ethnic or religious identity, figure in this, as do hope and trust, for which a properly functioning democracy and the rule of law, whereby all citizens are treated equally, are relevant. Dominique Moïsi has mapped the world in these terms in his writing on the geopolitics of emotion.[77]

The dominant expectation in the Western world around the turn of this century was that – after the disappearance of the threat from the Communist bloc following its collapse in 1989 – the growth of a worldwide network of open societies and free-market economies would be sufficient to guarantee lasting peace. Some predicted that a 'clash of civilisations' would stand in the way of that, but the simplifying geographic assignment of characteristics of civilisation to countries and peoples was unconvincing. After all, almost everywhere different types of 'civilisation' which rival groups seek to join are to be found within one and the same society. A far more important question seems to be what does or does not hold a society together, in other words what determines the degree of social cohesion in a society. As already mentioned, Dominique Moïsi has referred to the role emotions play in international relations, with hope, humiliation and fear as the principal affects. A carefully thought-out, realistic security strategy must take these emotions into account and simultaneously help to create the conditions in which emotions can play a positive role in promoting peace and security. As Moïsi notes, hope comes from stable mutual trust.[78]

Feelings of hope, humiliation and fear dominate in various parts of the world. The conflicts arising from these feelings - within, but certainly also between societies - explain revolutionary movements, rebellions and wars – the feeling of humiliation that dominates in the Arab world, for example. The Russian desire to regain its legitimate place in history is an important factor in Putinism.[79] Chinese nationalism is also rooted in a deep sense of being short-changed.[80] In interaction with shifting power relations, the geopolitics of emotions is leading to a 'mixed revival of geopolitics in Europe'.[81] The underlying social micromechanisms must be given a place in a better underpinning of security policy. This is one of the reasons why the culture of hope in Western Europe has given way to widespread, but also often diffuse fears. It has frequently been observed that the manifestation of violent international conflicts has changed, but the nature of the associated affective dimension is not sufficiently recognised.[82]

These patterns translate into an altered experience of security: the absence of human security reinforces frustrations and fears; insufficient improvement in the functioning of governments constrains mutual trust. The rise of nationalist regimes that oppress or persecute other groups leads directly to insecurity. The rise of terrorist (in other words, frightening) movements, sometimes in reaction to repressive regimes, but often in fact opposed to democracies, further undermines the feelings of security and causes migration movements, which in turn have their own disruptive effects. Terrorist organisations, piracy on supply routes, off the Somali coast for example, and forms of aggression coming from fragile states also undermine an aspect of security that we describe as flow security.

There are also mixed forms of conflict, as we have seen in the eastern regions of Ukraine, in which national groups and foreign – in this case Russian – soldiers mingle to the point of becoming indistinguishable. The most important point in that context is that besides the 'classical' socio-economic causes of armed conflicts, the emotional swings and confrontations have a powerful knock-on effect in terms of a willingness to engage in armed conflict, and hence genuine insecurity. Those who flee situations in which they feel unsafe and hopeless arrive as asylum migrants in countries in which their presence is portrayed by some as a source of insecurity, although in the Netherlands they are not suspected of crimes more frequently than native Dutch persons with similar demographic and socio-economic profiles.[83] In other words, in this day and age violence and insecurity are no longer solely the consequence of decisions by governments to take up arms, but also the consequence of a dynamic of popular emotions, both within a state and across borders.

These changes in the nature and perception of conflicts point to an important distinction compared with earlier wars driven by geopolitical motives. Formerly – especially in the First World War – populations and armies were persuaded relatively easily to obey their commanders by existing motives of national pride and loyalty ('in defence of the fatherland'), even when they were sent to sacrifice themselves in the trenches; this loyal following of orders was enforced with military discipline. In a time of extensive communication networks,[84] the leadership has become even more dependent on the intensity with which strong emotions can be stirred in broad sections of the population. The 'total war' conducted by the Third

Reich was the product of the unprecedentedly intensive ideologisation of the German people. In contrast, the British and American war efforts were driven, far more than during the First World War, by the realisation that it was not only their interests but also their fundamental values that were at stake. The incapacity of the American political leadership of the time to present the Vietnam War as a national calling was a turning point in American foreign policy; only in reaction to the attacks on 9/11, and then only for a few years, could the American public again be mentally mobilised for a distant war. The strength of Da'esh lies not in superior weaponry, but in the total subjugation of the fighters to a goal that is presented as super-human, is supported by modern communication technologies and plays on both frustrations and religious codes.[85] In a commercially-oriented culture like that of Western Europe, such extreme movements appear all the more frightening.

Populations stir themselves if regimes expose them to suffering and death for a military objective they do not share. On the other hand, the suggestibility arising from existing feelings of frustration and fear itself can be a potential source of violence. Especially in the digital era – with all of the possibilities it offers to manipulate the supply of information with the help of hackers or troll factories – this 'horizontalisation' is in no way reassuring. It is therefore necessary, both in this country and elsewhere, to incorporate the experiences and feelings of the citizens in a broad and thorough reorientation of security policy (for example, among the populations of unstable societies on the other side of the Mediterranean). The concern for state security focused on defence of the territory must be linked to other dimensions of security.

This is manifested to a large extent at the crossroads – in terms of connections and cultures – of the three continents of the old world: the Near East. The destabilising factors include (a) the international feeling of inequality in standards of living, such as the enormous growth of the population in regions such as North Africa without a corresponding development in terms of the economy, education and infrastructure; (b) a significant absence of European involvement in those domains in the immediate vicinity, in particular in the countries to the south and east of the Mediterranean; and (c) a military-political power vacuum in those same regions, with failing states and complicated 'new wars' that are totally different in character from the classical inter-state wars or civil wars.[86]

2.6 Challenges for Security Policy

The intensification of cross-border flows, the sharp increase in the number of non-state actors and the disappearance of national border controls within the EU underscore the complex interconnectedness of internal and external security. What does this mean for governments? They will be confronted more often with transnational (security) issues whose origin, course and consequences are difficult to oversee. Examples are the unexpected appearance of criminal networks with worldwide branches, but also the disruption of complex production chains and technical systems by sabotage or technical or human failure.[87] Highly developed, open societies appear to be relatively vulnerable to the threat from lone wolves or groups that turn against them and who exploit the blurring of the boundary between ‘internal’ and ‘external’. After all, the authorities are bound by their own statutory powers and international human-rights treaties. Their actions are under the microscope of the media, critical NGOs and the articulate general public. Furthermore, there is often less public tolerance for risks and uncertainties in those societies precisely because of the strong belief in the malleability of society and the high expectations for government intervention. A government that does not meet those expectations with visible security measures can quickly stir up feelings of fear and insecurity. Equally, however, a government that does meet them can reinforce the feelings of insecurity. If politicisation in society comes to overly dominate security policy itself, it can lead to a negative spiral of fear and mistrust, which could even threaten the open society itself.[88]

Prioritising

This presents the government with a number of major challenges. The first involves distinguishing and prioritising: how can it still rank the variety of claims to national and international security? In what areas are fundamental public interests at stake which require the government to play a public role (as director or otherwise), and in which are they not? Processes of securitisation are due in part to the expansion of the security agenda. It is therefore essential to guard against ‘normative overdemand’ on government.[89] Naturally, prioritisation is ultimately a question of making political and normative choices. But in the process of making those decisions, sound considerations, based on relevant knowledge and public deliberation, are essential for taking effective action and securing public support.

Governance

The second and connected challenge concerns the division of roles and coherent action by the relevant actors (governance). Globalisation, Europeanisation, socialisation and privatisation call for a rearrangement of powers, tasks and relationships among the various actors. Which interests can now be performed better by European or global institutions or civil-society organisations and which cannot? Where is joint action desirable, and what demands does it place on the growing number and variety of civil and military actors at social, national and supranational level? The ‘scaling up’ of security issues to European and global level and the (partial) privatisation of security functions often lead to uncertainty about who is responsible and legally and political liable for specific security issues, and under what circumstances. We see this, for example, in the international security and stabilisation missions carried out by multinational coalitions of countries, international organisations and military and civil actors. These missions blur the boundaries between the civil and military roles and powers, demand a high degree of mutual cooperation and constant coordination within and between nation states, ministries, civil-society organisations and other non-state or semi-state actors. They also call for a larger degree of interoperability of civil and military materiel capacities. There is a reason for the constant calls for a more ‘integrated approach’ to international security issues and missions. This wider range of international security challenges presents smaller countries with limited resources in terms of materiel with relatively major challenges, which is another factor that compels a rethinking of the priorities of security policy and resources needed to implement it.[90]

Diffuse Threats

The third challenge, finally, arises from the gradual expansion of the danger dimension of security. In addition to immanent threats, more diffuse, potential future vulnerabilities and uncertainties play an increasingly important role. The clear and present danger of the nuclear threat that was prominent in the security agenda during the Cold War has faded into the background somewhat, although the role of nuclear weapons has again risen higher on the security agenda in recent years with the 'return' of power politics. But that agenda also includes less obvious threats, ranging from global climate change and international terrorism to energy insecurity. In principle, the vulnerability of humans, society and the natural environment calls for a proactive response to uncertainties,[91] but where reactive and defensive military action makes way for a preventive, interventionist approach, as in the case of the invasion of Iraq in 2003, it creates tensions with the classical principles of national and international law, the fundamental rights of citizens and public support.[92] This also applies to a growing extent for various new preventive European and national policies in relation to counterterrorism, cyber security or tackling irregular migration.

Endnotes

1 Wijk, R. de & Toxopeus, R. (2005).

2 Renn, O. (2008).

3 Herington, J. (2015: 30).

4 Wijk, R. de (2015a); Guzzini, S. (2012).

5 Pelkmans, J. (1982: 3).

6 Daase, C. (2013); Hough, P. (2013).

7 Losman, D. (2001: 5); Hough, P. (2013: 157).

8 Lynn, B.C. (2014).

9 Brattberg, E. & Hamilton, D.S. (2014: XVI).

10 The World Economic Forum's Global Risk Report 2013 identified no fewer than 50 global risks, most of which are interconnected, or at least correlated. The European Central Bank (ECB) has observed on the basis of 43 volatility shocks on the international stock markets that globalisation brings with it new opportunities, but also acute global uncertainty and risk aversion (WRR 2013: 157–8).

11 Adelman, M.A. (1972).

12 Gehem, M. et al. (2015: 58).

13 WRR (2015).

14 Martinage, R. (2015).

15 Stehfest, E. et al. (2014); HCSS (2014).

16 Missiroli, A. et al. (2014).

17 Brattberg, E. & Hamilton, D.S. (2014: XI).

18 Brattberg, E. & Hamilton, D.S. (2014: 22).

19 IOM. (2008: 30).

20 Hough, P. (2013); Frerks, G. & Klein Goldewijk, B. (2007: 29–30).

21 Kaldor, M. (2014: 86–87).

22 UNDP. (1994: 22–23).

23 Soon after the publication of the Human Development Report, human security was enthusiastically propagated by the United Nations Development Programme (UNDP) and other UN institutions such as the World Health Organisation (WHO). Although the Security Council should normally confine itself to acute crises, in 2000 a non-military security resolution was adopted for the first time during a debate on HIV/Aids. The American government under President Clinton was one of the driving forces behind this expansion of the security agenda at state level. Inspired by Joseph Nye's scientific publications about the significance of soft power and knowledge as an important non-military dimension of state power, the American national security strategy in 1994 explicitly stated that not all security risks are military in nature. Transnational phenomena such as terrorism, environmental damage, rapid population growth and growing refugee streams would also have security implications for the US in the short and longer term (Hough 2013: 15).

24 Hough, P. (2013: 10).

25 In its report on humanitarian interventions in 2000, the Advisory Council on International Affairs (AIV) referred to the confusion surrounding this term. It called for a clear distinction to be made between (1) providing assistance without the consent of the relevant country in order to alleviate acute emergencies that constitute a threat to the lives of large numbers of people (humanitarian emergency assistance); (2) authorisation by the Security Council of the use of force on the basis of chapter VII of the UN Charter in response to situations involving large-scale violations of human rights (authorised humanitarian intervention); and (3) intervention by a state or group of states, involving the use or threat of force, on the territory of another country in response to grave, large-scale violations of human rights that are taking place there, without the prior authorisation of the Security Council (humanitarian interventions) (AIV 2000: 6).

26 Pingeot, L. & Obenland, W. (2014: 17).

27 R2P contains three components: a responsibility to prevent, a responsibility to respond and a responsibility to reconstruct. This principle was invoked for military interventions in Darfur, the Democratic Republic of Congo, Ivory Coast, Mali, Libya and against Da'esh.

28 AIV. (2010: 8); Seybolt, T.B. (2008: 2).

29 Weller, E. (2014); Seybolt, T.B. (2008: 14); Pingeot, L. & Obenland, W. (2014: 39).

30 Pingeot and Obenland (2014: 33) point out that the Canadian R2P report in 2001 and the official UN document in 2005 have a different status, but are often not clearly distinguished.

31 Weller, E. (2014: 7).

32 Christou, G. (2014: 369).

33 Terpstra, N. et al. (2014: 14).

34 See also the report of the Human Security Study Group (2016), which called for a 'second generation human security approach' for the EU as an alternative to the War on Terror and 'realistic' geopolitics.

35 AIV. (2010: 8).

36 Wilde, J. de (2008: 229).

37 Wilde, J. de (2008: 237).

38 Martin, M., Kaldor, M. & Serro, N. (2013); Liotta, P.H. & Owen, T. (2006: 39); Weller, E. (2014: 9).

39 Naturally, the notion that state sovereignty and human rights are two sides of the same coin is not new. It is also reflected in the UN Charter. The AIV (2010: 21), for example, refers to the Plakkaat van Verlatinghe in 1581 (the Act of Abjuration; the declaration of independence from Spain by some provinces of the Netherlands), Grotius' treatise on law and the American Declaration of Independence as sources which take as their basic principle that state sovereignty assumes the obligation to safeguard the interests of individual citizens.

40 Weller, E. (2014: 9).

41 https://news.un.org/en/story/2017/07/561142-humansecurity-approach-central-achieving-sustainable-development-un-officials

42 UNDP (1994).

43 Hough, P. (2013: 13).

44 Cf. EU Global Strategy (2016).

45 AIV. (2016).

46 AIV, Security and Stability in Northern Africa, May 2016, p. 44.

47 AIV. (2016: 44).

48 AIV. (2016: 46).

49 Buzan, B. (1991).

50 See Article 6 of the North Atlantic Treaty (1949):
For the purpose of Article 5, an armed attack on one or more of the Parties is deemed to include an armed attack:

- on the territory of any of the Parties in Europe or North America, (…) on the Algerian Departments of France, on the territory of or on the Islands under the jurisdiction of any of the Parties in the North Atlantic area north of the Tropic of Cancer;

- on the forces, vessels, or aircraft of any of the Parties, when in or over these territories or any other area in Europe in which occupation forces of any of the Parties were stationed on the date when the Treaty entered into force or the Mediterranean Sea or the North Atlantic area north of the Tropic of Cancer.

51 Simma, B. & Müller, A.T. (2012: 137–38).

52 See Article 97(1) of the Constitution of the Kingdom the Netherlands: "There shall be armed forces for the defence and protection of the interests of the Kingdom, and in order to maintain and promote the international legal order." See also Article 100(1): "The government shall inform the States General in advance if the armed forces are to be deployed or made available to maintain or promote the international legal order. This shall include the provision of humanitarian aid in the event of armed conflict."

53 Parliamentary Documents II 2002–2003, 28 880, no. 22, p. 6.

54 The term is rightly criticised as being undifferentiated and departing from the premise of a dichotomy between state and non-state actors. For a more extensive and nuanced explanation of the many guises of these actors, see, among others, HCSS 2014; Kaldor and Rangelov 2014. In its advisory report on the interaction between actors in the international system, the AIV distinguishes four categories of actors: bilateral, multilateral, civil and private. Civil actors are civil-society organisations and research institutes that occupy a position in society between the household, the state and the private sector (AIV 2013: 8).

55 HCSS. (2014: 141–147).

56 Dennis Blair, Director of National Intelligence, in 2010 explained this as follows: "No dominant adversary faces the United States that threatens our existence with military force. Rather, the complexity of the issues and multiplicity of actors – both state and non-state increasingly constitutes one of our biggest challenges." Cited in: De Graaf 2012a: 16.

57 Graaf, B. de (2012a: 16).

58 As Realuyo puts it: "Illicit networks seek to navigate, infiltrate, and/or dominate global supply chains to further their activities and enhance their power. They actually thrive in open societies with the free flow of goods, people, and capital, and information for their clients. Illicit actors utilize and even seek to control or co-opt supply chains around the world to facilitate the movement of "bad people and bad things" such as drugs, guns and counterfeit goods." (Realuyo 2013: 2).

59 Missiroli, A. et al. (2014: 44; 60); Rood, J. & Dinissen, R. (2013: 103); HCSS (2014: 149); Graaf, B. de (2012a).

60 Pilbeam, B. (2015a, b); Abrahamsen, R. & Williams, M. (2010).

61 Deitelhoff, N. & Fischer-Lescano, A. (2013: 63); Chakrabarti, S. (2014: 506).

62 The reservation also applies to this development that it is not entirely new, but its pace has greatly accelerated because of privatisation and globalisation. Machiavelli was already very critical of mercenary armies, and mercenaries were also relied on in the major European conflicts in the eighteenth century, the American War of Independence, the American Civil War. Hiring private compa-

nies for domestic security was also not uncommon in the more developed Western countries up until the nineteenth century. The American company Pinkerton, which started as a detective agency in 1850, also supplied armed guards and strike-breakers for industrial companies.

63 Deitelhoff, N. & Fischer-Lescano, A. (2013: 56).

64 Governments also increasingly depend on threat analyses, research and the development of new security technologies by private companies. According to estimates in the Small Arms Survey Yearbook 2011 based on reports from 70 countries, between 19.5 and 25.5 million people were working worldwide in the private security industry, which had an estimated annual turnover of 218 billion dollars and was operating in practically every country in the world (Small Arms Survey Yearbook 2011: 101–102; Rood 2013: 110; Chakrabarti 2014: 509). The annual growth of this sector is estimated at over 7%.

65 Andrews-Speed, P. et al. (2012: 46–65).

66 Buzan, B. et al. (1998); see also Chap. 1.

67 Hagmann, J. (2015).

68 Malik, S. (2015a, b).

69 Hagmann, J. (2015: 39).

70 De Wijk (2015a: 16–17) refers in this context to the 'largely disappeared' distinction between high politics (the security and ultimate survival of the state) and low politics (the functioning and preservation of the services of the welfare state).

71 Bailes, A.J. (2013); Daase, C. (2013).

72 Cf. WRR (2011).

73 See also the Catalogue of National Operations [Catalogus Nationale Operaties] (2016) with an extensive overview of the tasks the defence forces carry out during national operations: https://www.defence.nl/onderwerpen/taken-in-nederland/downloads/brochures/2016/09/09/catalogus-national-operaties-pdf-versie [Dutch only].

74 Graaf, B. de (2012d: 55).

75 Bourbeau, Ph. (2015: 10).

76 "The ongoing process of globalisation has taken us into a hyper-connected world. The digital and ICT revolution have raised connectivity to a new level. This has brought about a global connectedness that is transforming our social, political, economic, and cultural-ideational world profoundly. It affects the human sense of time and place: the world is both shrinking and expanding, familiar and alien, highly complex and unpredictable. It renders the global and the local inseparable." Strategic Plan Asser Institute, http://www.asser.nl, p. 8.

77 Moïsi, D. (2009; 2015).

78 Moïsi, D. (2015: 67).

79 Foreign Affairs (June 2016).

80 Wijk, R. de (2015a: 189).

81 Guzzini, S. (2013).
82 Hutchison, E. & Bleiker, R. (2014: 491–515).
83 WRR (2015: 15–21).
84 Castells, M. (2009).
85 Luizard, Pierre-Jean (2015).
86 Kaldor, M. (1998); Pilbeam, B. (2015a, b).
87 Eriksson, J. & Rhinard, M. (2009: 247).
88 WRR (2006); Rood, J. (2014: 134).
89 Daase, C. (2013).
90 Bailes, A.J. et al. (2013: 37); WRR (2010a, b).
91 WRR (2008); Daase, C. (2013: 32–33).
92 Ducheine, P. (2008).

Chapter 3
Human Security and Flow Security in Dutch Security Policy

Ernst Hirsch Ballin, Huub Dijstelbloem, and Peter de Goede

3.1 Introduction

The expansion of the concept of security has also had an impact on security policy in the Netherlands, where security now encompasses far more than the military defence of the country's own territory against hostile armies. The concern for human security in fragile situations elsewhere in the world and the economic relevance of flow security are relatively new focus areas in Dutch security policy. Accordingly, this chapter provides a further analysis of the integrated approach to security and development on the one hand, and security and the economy on the other, with an emphasis on the relevant policy efforts of the Netherlands.

Security policy in the Netherlands has gradually departed from the view that security is a question of protecting the national territory against external forces. After the Second World War – in the context of the 'Pax Americana' – the policy of neutrality was abandoned and the Netherlands joined the North Atlantic Treaty Organisation, whose purpose was the *collective* defence of the alliance's entire territory against the external threat from the Soviet Union. When the Cold War ended – and it was assumed that the threat had passed – targeted interventions and preventive actions in other parts of the world also entered the realm of security policy. The focus of thinking about security shifted from the Cold War dynamic to crisis management in fragile states, human rights and development (human security). The agenda for development cooperation was bound up with operations involving deployment of the armed forces. That shift is illustrated by the addition, in 2000, of 'the maintenance and promotion of the international legal order' to the constitutional tasks of the armed forces, not only as an expression of international solidarity, but also as a matter of enlightened self-interest. Particularly after the terrorist attacks of 9/11, we felt less safe and there was a growing

E. Hirsch Ballin (✉) · H. Dijstelbloem · P. de Goede
WRR, Scientific Council for Government Policy, The Hague, Zuid-Holland, The Netherlands
e-mail: ballin@tilburguniversity.edu; h.o.dijstelbloem@uva.nl; peter.goede@minbzk.nl

E. Hirsch Ballin et al. (eds.), *Security in an Interconnected World*,
Research for Policy, https://doi.org/10.1007/978-3-030-37606-2_3

realisation that our own security was closely entwined with the consequences of *human insecurity* elsewhere in the world.

This integrated approach to security and development was in fact introduced in the context of a substantial reduction in government spending on defence. Following the end of the Cold War the optimistic view prevailed that military conflicts would pass this country by, and after the financial and economic crisis of 2008 there was a need to cut government spending, which also had a negative effect on spending on – conflict prevention through – development cooperation (Sect. 3.2).

Today, in 2017, the realisation of the international interconnectedness of our security situation is stronger than ever and not just because of the escalating streams of migrants in recent years, which have also led to border security assuming a central place in the national and European security agenda. We live in a network society in which production, consumption and circulation are organised on a global scale. Capital, energy, raw materials and information flows are organised in continental and intercontinental networks and transcend traditional location-bound connections. The 'space of flows'[1] creates a new international division of labour and capital which affects the entire planet, but does so asymmetrically through either inclusion in or exclusion from the global economy. Not in the simplistic form of a centre and a periphery or in the distinction between 'North' and 'South', because there are multiple centres and multiple peripheries. Furthermore, North and South are internally so differentiated that categorising according to those concepts is pointless.

Within the global web, nation states or associations of states (such as the EU) compete with each other to protect the interests of their enterprises and citizens. Their security is closely connected with the worldwide networks and any disruptions to them. States are therefore increasingly active in addressing issues relating to economic security (flow security). Energy and climate, raw materials and the cyber domain have therefore also become part of a socialised security agenda in the Netherlands. In other words, security is also a question of the variety of connections on which our economy depends (Sect. 3.3).

3.2 Human Security: The Integrated Approach to Security and Development

3.2.1 Policy Towards Fragile States

The number of intra-state conflicts has grown since 1990, particularly in the least developed countries.[2] In the course of that decade a wider consensus emerged that international action in the form of external intervention in fragile – or failing – states was justified, provided it was backed by a UN mandate. Definitions of what constitutes a fragile state varied[3] from countries with ineffective or weak governance or countries whose central government had lost authority over some or all of the territory to countries where the government could no longer guarantee the security of its own citizens. Solidarity with the local population, providing human security, protecting the international legal order and combating terrorism and international crime could all be grounds for intervention (including military intervention) in fragile states. There were also references to 'enlightened' self-interest, since the negative effects of conflicts, such as crime, refugee streams and terrorism, could spill over into neighbouring countries, to an entire region and to the West and also undermine stability there. Civil liberties are curtailed in many fragile states (see Fig. 3.1). The idea is that stabilising fragile states is ultimately also beneficial for civil liberties in those countries, and hence for their development.

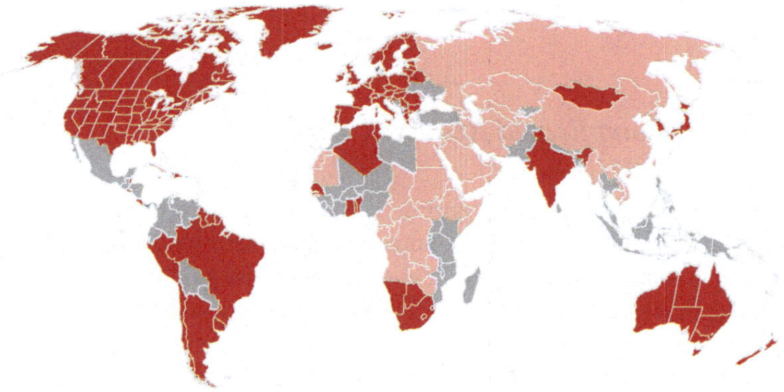

Fig. 3.1 Public freedom in the world 2014. (Source: Clingendael, Strategic Monitor 2014, *Een wankele wereldorde* [A shaky world order], 2014: 48)

The growing attention for fragile states had three consequences. The first was that it brought an end to the former philosophy that development aid only really helped if countries had attained a certain degree of stability, security and 'sound' – read: efficient and honest – governance. That view had prompted many donor countries to attach great weight to the criterion of good governance in deciding whether to enter into, break off or revise aid relationships – in the same way as they had done earlier with the human rights criterion.[4] The second consequence was that the complex relationship between security, stability, reconstruction and development rose higher on the agendas of international organisations and Western donor countries. The third was that political, civil and military activities increasingly started to converge in the practice of intervention and crisis management operations.

The emphases of Dutch policy towards fragile states have differed as the policy has evolved since 2005. In the policy document *Wederopbouw na gewapend conflict* [Reconstruction after armed conflict] (2005), reconstruction in fragile states was described as a 'complex process that demands an integrated approach' and was said to have three dimensions: consolidation of peace and security, rehabilitation of public administration and restoration of basic services and employment. At national level, the coherent, integrated approach that was advocated implied that the relevant ministries (Foreign Affairs, Defence, Economic Affairs, Justice, and Interior and Kingdom Relations) would select a small number of countries in three priority regions – the Great Lakes and the Horn of Africa, the Western Balkans and Afghanistan. They would also have to draft joint strategies for specific countries, jointly report on the financial resources they were reserving and coordinate their respective funding. In addition, 'a consultative structure [would] be created to further improve the cooperation'.[5]

3.2.2 The Complicated Reality: Uruzgan

The Dutch military presence during NATO's Enduring Freedom operation in Afghanistan between 2002 and 2010 demonstrated increasingly painfully how complicated interventions and reconstruction operations in fragile states can be in practice. In Uruzgan (2005–2010), what started as a display of solidarity with the allied fight against international terrorism shifted to an operation with the ambition of improving stability and security in the province by winning more support for the Afghan authorities among the local population. A lot was at stake politically with the mission in Uruzgan. The Ministry of Defence, and within it the Operations Directorate, regarded the Netherlands' occupation of a position as a lead nation as a litmus test for the capacity of the Dutch armed forces to make a visible contribution to complex military operations.[6] One of the factors in this was the fervent desire to remove the blot on the record of the armed forces caused by 'Srebrenica' and to show that despite all the spending cuts the armed forces were actually performing an important public duty.[7]

Grandia shows in her thesis that the context in which the Netherlands decided to send troops to Uruzgan was determined to a large extent by the political considerations of membership of the alliance. However, the initiative for the mission, the initial exploratory negotiations with the British and the Canadians, and even the planning, preconditions and location of the Dutch mission initially came exclusively from the military and were not primarily dictated by pressure from the NATO partners.[8] It was the Defence Staff itself that proposed to the officials at the Ministry of Foreign Affairs and their own minister that a 1000 troops should be deployed, in the expectation that a proposal for a larger force would not be politically feasible. Once that number began circulating in political circles, there was no turning back, even when it became clear that the security situation in Uruzgan was deteriorating.[9] Accordingly, the so-called Article 100 letter to parliament referred to a stabilisation operation for 2 years with a maximum of 1200 troops and a budget of 320 million euro. The so-called Assessment Framework and the Article 100 procedure for the decision on deployment of Dutch troops proved to be no guarantee of a diligent decision-making process by the cabinet or of the timely provision of information to the House of Representatives. After the military planners had already anticipated the politicians, the military and political decision-making processes then quietly converged, with the result that the political primacy was weakened and the deployment of troops came to appear inevitable to those involved. Nor had the Steering Group formulated a strategy for the deployment of troops with predetermined goals and resources and a further interpretation of the integrated approach of defence, development and governance as promised to the House of Representatives. The Ministry of Defence finally drafted a master plan, but it was only published just before the first troops arrived in Afghanistan, while an estimate of the situation of civilians in Uruzgan prepared by the Dutch embassy in Kabul was only published after the troops were already in the country.

Neither document was incorporated into an overarching strategic document so the goal of the mission remained diffuse.[10] According to Amersfoort, 'on balance' the Uruzgan Task Force was deployed 'without any underlying military strategy or campaign plan'. He referred to 'the demise of military strategy'.[11]

Under pressure from the House of Representatives, the media and public opinion, the Dutch government was anxious to distance itself from the actions of the Americans, and to a lesser extent the British and the French, because of their focus on 'counterinsurgency': the suppression of insurgencies and the military elimination of Taliban and Al Qaeda fighters.[12] The government carefully avoided using the term counterinsurgency, or the term war. The Netherlands would make a 'unique contribution' in the form of the 'Dutch Approach'.[13] Later, following the Canadian example, this approach was rechristened the '3D approach' (for diplomacy, development and defence) and the 'integrated approach', in which the prevention of conflicts is also an important facet. The aim was 'to act with respect for the population, knowledge of religion, local habits and customs and with as little aggression as possible'.[14] The Netherlands would address the underlying causes of instability, for example by rooting out the cultivation and trafficking of drugs, offering alternative means of subsistence, reforming the security sector and combating poverty.[15] The civil- and development-oriented dimensions of the mission thus attracted more and more of parliament's attention and the pressure increased to involve NGOs in the mission from an early stage. An example of how remote the reality of the mission was from the political discussion in the Netherlands was the debate that was conducted about whether Uruzgan was a 'reconstruction mission' or a 'combat mission' when the complex reality on the ground meant that in various places reconstruction and fighting were proceeding simultaneously.[16]

In the run-up to the extension of the mission at the end of 2007 initially optimistic reports about the progress being made in finding the optimal combination of military resources, development aid and political and diplomatic instruments steadily made way for more nuanced views and revisions because of 'the complicated reality [which] compels realism and pragmatism and an increasingly nuanced and more complex view of the conflict in Uruzgan'.[17] There was every reason for this change. Maintaining security and stability, specified as a prerequisite for establishing good governance and socio-economic development, proved impossible to realise with the limited number of Dutch troops. There were not enough transport helicopters and engineers and medical personnel to provide support. The mobility of the troops was also severely hampered by improvised explosive devices (IEDs). It was therefore gradually decided that the Dutch troops would secure stable zones ('ink stains') around three major population centres in the province. Within these areas, roughly half of the civilian population of Uruzgan could count on – temporary – protection. Outside them the rebels had free rein, which meant there was no chance of winning the 'hearts and minds' of the civilian population.[18] Nor was any progress made in achieving the ambition of training the Afghan army and police to assume some tasks themselves, for example taking charge of anti-drug operations. The limited capabilities of the Afghans was a factor in this, but also divisions among the ISAF partners about how to deal with the local tribal leaders and an unwillingness on the part of the government in Kabul and the local authorities to tackle corruption, incompetence and nepotism.[19]

With hindsight it was observed that the mission's objectives as set out in the original Article 100 letters were 'ambitious and impressionistic'.[20] Until October 2007, the policy documents sent to the House of Representatives reflected an optimistic belief in the malleability of the situation, despite the worrying developments that were occurring in terms of security. The Netherlands was said to be contributing to the reconstruction and transition of a failing state into a functioning constitutional parliamentary democracy on the basis of lengthy joint efforts by the international community. The more successful the Afghan government was in strengthening its own authority, the more the International Security Assistance Force (ISAF) would become redundant. At the same time, no realistic and concrete objectives had been formulated in advance and there were no indicators to measure results nor was there any underpinning of the integrated approach. Nor was careful thought given in advance to how the UN's call for civilians to be protected would be translated into guidelines for strategies, mandates and reports concerning their protection.[21]

Despite – or perhaps because of – these flaws, the Dutch actors on the ground developed a flexible, pragmatic and practical approach in which the various ministries, civil and military actors and NGOs achieved a steadily greater level of coherence in their actions at operational level in a constantly changing environment. The Uruzgan Task Force made 'progress in finding the best combination of military resources, development cooperation and political and diplomatic instruments'.[22] At the same time, however, it became increasingly clear that numerous developments were occurring over which the Netherlands, as a small actor facing problems of capacity in the province of Uruzgan, had no control whatsoever.[23]

3.2.3 Adjustment of the Policy

Against the backdrop of the dramatic Dutch experiences in Uruzgan and the continuing debate about the characteristics, causes and consequences of fragility, Dutch policy towards fragile states was revised and given a better strategic underpinning.

Veiligheid en ontwikkeling in fragiele staten [Security and development in fragile states] (2008), a document that fleshed out the policy letter *Een zaak van iedereen* [Our common concern] (2007), outlined the multidimensional nature of the problems of fragile states and the need to focus on civil operations and to form multilateral coalitions wherever possible. The report stated that conflict prevention through diplomacy and political efforts coordinated by the UN would occupy a prominent position in the strategy, for example via early warning and mediation. The Netherlands should raise simmering conflicts in multilateral forums and would assist civil-society organisations in securing independent information about developments in countries and in carrying out projects designed to address the underlying causes of fragility. 'Islands of stability' could also receive assistance in preventing the expansion of conflicts. The Ministry of Foreign Affairs would also enhance its own policymaking capacities by creating a flexible Fragility and Peacebuilding Unit to operationalise the policy towards fragile states. The unit would work closely with embassies, other ministries, NGOs, international organisations and experts.[24]

Meanwhile, prompted by a debate about the 3D approach in the Senate and ahead of the publication of the so-called *Strategische Verkenningen* [Strategic Defence Review] by the Ministry of Defence, the government asked the AIV to write an advisory report on the policy on fragile states. The request reflected the struggle with the experiences in Afghanistan. The AIV was asked to say how the political, military and development objectives of crisis management operations relate to each other in theory and practice, to what extent those objectives can be integrated in a single coherent approach and how an integrated approach should ideally be put into practice.[25]

In its report published in March 2009, the AIV analysed the many dilemmas posed by crisis interventions and reconstruction. The Council warned the government to display 'sobriety and moderation' by formulating limited, attainable objectives, which should also be discussed more thoroughly in advance at the political level. The Council argued that the chance of success and the quantifiability of successes are crucial for preserving the essential political and public support for missions. It was understandable that the government is mainly inclined to highlight the positive aspects of peacekeeping operations and is keen to 'keep the public's spirits up', said the Council, but, it warned, "all this can have a fatal impact on society's indispensable faith in a good outcome and erode public support if reality proves to be more intractable".[26] The AIV also observed that the limited capacity to deploy civil servants, the limited capabilities of the armed forces and a lack of knowledge of the situation on the ground are major obstacles.

At the end of 2011, the *New Deal for Engagement in Fragile States (2012–2015)* was adopted in Busan under the chairmanship of the Netherlands. The agreement underscored the need to strengthen the policy on fragile states. An updating of Dutch policy on fragile states followed,[27] in which the government specified five priorities: (1) human security, (2) a functioning legal order, (3) inclusive political processes, (4) a legitimate and capable government and (5) peace dividend: employment and basic services.[28]

The Department for Stability and Humanitarian Aid (DSH) of the Ministry of Foreign Affairs and the relevant embassies drew up Multi-year Strategic Plans for fragile states such as Rwanda, Afghanistan and South Sudan.[29] However, they did not do the same for the ring of unstable countries in Eastern Europe and in North and West Africa, despite these countries' strategic importance for the Netherlands and Europe. The DSH provides funding to five organisations (International Alert, Interpeace, International Crisis Group, Centre for Humanitarian Dialogue, International Centre for Transitional Justice) for activities relating to conflict prevention, conflict resolution and peacebuilding. Some of these activities involve early warning, some mediation and some peacebuilding. The organisations are strategic partners of the Ministry of Foreign Affairs and have been receiving non-earmarked grants since 2009 (some since 2011).[30]

3.2.4 An Integrated Approach?

Since 2008 the Dutch government has frequently referred to 'the integrated approach' in policy documents such as *Wederopbouw na gewapend conflict* [Reconstruction after armed conflict] (2005), *Security and development in fragile states* [Veiligheid en ontwikkeling in fragiele states] (2008), *Focusbrief Ontwikkelingssamenwerking* [Focus letter on Development Cooperation] (2011), *Internationale Veiligheidsstrategie* [International Security Strategy] (2013) and *Leidraad Geïntegreerde Benadering* [Guidelines for the Operationalisation of the Integrated Approach] (2014). The integrated approach is one of the six priorities in the International Security Strategy. In January 2014, for example, the Minister of Foreign Affairs wrote to the House of Representatives, with reference to the Coalition Agreement (2012), that the government supported proposals to develop a joint integrated approach at EU level: 'the Netherlands is seen within the EU as one of the drivers of the integrated approach, particularly because of the experience the Netherlands has gained with that approach in Afghanistan and which has been translated into policy within organisations such as NATO.' But, according to the minister, the EU's ambitions did not go far enough. Opportunities are being missed to genuinely end the compartmentalisation of the relationship between security and development on the basis of joint civil-military planning and financial programming and in combination with the New Deal for Engagement in Fragile States.[31] The letter to the House of Representatives in November 2014 concerning the International Security Strategy also refers to the desire 'to maintain and further expand the Netherlands' progressive role in relation to the integrated approach'.[32,33] The integrated approach had in fact already been embraced in the context of the UN, EU and NATO.[34] The establishment of the international Civil-Military Cooperation Centre of Excellence (based in the Netherlands) testifies to the growing interest in civil-military cooperation in complex crisis situations and the realisation that conventional kinetic operations (with the use of force) – particularly in urbanised environments – increasingly interfere with non-kinetic measures intended to win the support of the population, for example.[35]

The references to an integrated approach to fragile states and zones of conflict and the desire to further strengthen it are inseparable from the mixed experiences during the Dutch mission in Uruzgan. The relevant players on the ground increasingly came to realise that the objectives of the mission left a lot of room for different interpretations. The military initially regarded the primary goal of the mission as being to defeat the Taliban and create security with a strategy of counterinsurgency.[36] The civil actors, including many diplomats and representatives of national and international NGOs on the spot, were principally concerned with improving the security of the local population as part of a wider development strategy. The NGOs in particular distanced themselves from military operations against the insurgents, since those operations not only harmed their credibility as neutral aid workers, but could also stand in the way of longer-term political solutions. After all, the best chance of them succeeding was if the grievances of the widest possible section of the population, including those of the insurgents and the (former) rulers, were removed. Particularly in the last 2 years of the mission in Uruzgan (2008–2010), this realisation of the complexity of the situation and the diverse viewpoints, goals and interests prompted those concerned to intensify their efforts to bring greater coherence and synergy to their joint operations. Mutual trust grew and there was tighter coordination and a sharper focus and greater consensus on the goals of the mission. This is a process that creates opportunities for a longer-term approach that takes greater account of the underlying causes of conflicts and of the wider range of civil as well as military instruments and actors. It also offers opportunities to gain wider public support and improve cooperation at international level.[37]

There were, however, a number of weaknesses in the implementation of the integrated approach by the Netherlands. First, improved coordination and closer cooperation between the actors are not enough if their objectives for the mission vary too greatly and there is no overarching analysis that is properly underpinned and enjoys political support. What are the goals to be achieved and the resources to be used in the specific context? What dilemmas are expected to arise? How can the civilian population best be protected, and at what cost? These analytical and substantive questions have to be explored first, before the added value of improved coordination and closer cooperation is considered, because the answers to them determine whether, and if so, which of the various instruments and actors in the extensive toolbox (defence, development aid, diplomacy, police, justice, trade) should be used. The authors of the *Leidraad Geïntegreerde Benadering* recognised the importance of an overarching analysis, customisation and the proper embedding of the 3D approach in international frameworks, but at the same time seemed to regard the approach primarily as an organisational process of coordination at senior civil-service level in the Missions and Operations Steering Group and of 'communication policy' by ministerial spokespersons rather than as a political task.[38] It is significant that in an international comparative survey interested German officials described the Dutch version of the integrated approach as being based on the polder model, in which consensus, cooperation and pragmatic experimentation are central features. "Assessments that things could run better quickly translate into a 'let's organize it' attitude rather than the initiation of lengthy strategic discussions".[39]

With the pragmatic polder model, material and bureaucratic interests of the individual ministries quickly gain the upper hand during negotiations. The result can then be that political decision-making is confined to the modalities of the military contribution to the missions via the Assessment Framework and the Article 100 letters.[40] The real political and public debates about the feasibility of the specific goals, the risks, the public support, the modalities and the available (civil) resources then come too late and are not exhaustive. What then remains for parliament is little more than 'micromanagement' of the military operations. Ministers then depend mainly on civil servants and military officers in the field for day-to-day decision-making and implementation and cannot fall back on guidelines concerning risks and the attainment of goals.[41]

In the second place, the integrated approach also ignores the reality of the dilemmas and conflicting values that can arise from differing ideological or political visions of the underlying causes of fragility and the potential for change. In other words, there are limits to how coherent action can be. De Coning and Friis refer to an analysis of no fewer than 336 peacekeeping operations that were undertaken by Germany, the Netherlands, Norway and the United Kingdom in the 1990s. Every one of them suffered from a lack of coherence (a 'strategic deficit'). In 55% of cases there was no connection whatsoever to an overriding strategy for the country concerned. If the underlying values, interests and mandates are inherently contradictory, there are limits to the benefits that can be generated by strengthening the procedural coherence. It is then better to recognise the impossibility of achieving complete coherence or to aim for limited coherence by accepting sub-optimal solutions. As De Coning and Friis demonstrate, there are inherent tensions between the long-term impact and the short-term results of interventions, between the values and mandates of the various actors and between the interests of the intervening actors and the local population. In practice, therefore, there is also far less room for coherence than is assumed by many civil servants and politicians who advocate the integrated approach. That realisation imposes a need for realism and moderation.[42]

A third weakness is that the substantive ambitions of the integrated approach (addressing the underlying causes of instability, preventing new conflicts and contributing to lasting solutions that are supported by the local population) conflict with the public and political support at home. Effective contributions demand a commitment for decades, but in the post-Afghanistan era the public support for 'open-ended' military interventions has greatly diminished. There is also less political support for such complex operations than there used to be.[43]

3.3 Flow Security: The Integrated Approach to Security and the Economy

The global network society is characterised by worldwide flows of capital, labour and information. That introduces economic security as a relatively new concept in the Dutch policy context.[44] In the narrow sense of security as the possession of the economic resources needed to guarantee national security, however, it has never been absent. Foreign and security policy are permeated with the realisation that the survival of a country like the Netherlands, with an open economy oriented towards trade and distribution, stands or falls with a safe and predictable environment. Economic security has therefore traditionally been connected with the desire for a stable international legal order, free trade, open transport routes and guaranteed access to European markets.

Economic security was explicitly incorporated in policy for the first time in the *National Security Strategy* (NSS) in 2007 as one of the five 'vital national interests': territorial, ecological and physical security, and social and political stability.[45] In the National Security Strategy, economic security turns mainly on 'the ability of the Netherlands to function without disruption as an efficient and effective economy. Economic security could be impaired, for example, if trade with an important foreign partner were to disappear'.[46] Examples of specific threats to economic security that are mentioned are extreme scarcity of energy and raw materials.

The *International Security Strategy* (ISS) in 2013 and the policy letter *International Security. Turbulent Times in Unstable Surroundings* in 2014 reflected a growing awareness of the interconnectedness of internal and external security and the vulnerability of the Dutch economy to external threats. Economic security is regarded as a separate security interest, in addition to the defence of allied territory and a properly functioning legal order. In contrast to the National Security Strategy, the International Security Strategy does not give a definition of economic security, but explicitly opts for an interpretation that – without explicitly using the term – corresponds with the broader concept of flow security. On the one hand, it refers to the importance of preventing undesirable interruptions of flows and production chains. On the other, it discusses the need to protect strategic economic sectors and vital infrastructure. Threats referred to include piracy, cyber attacks, (cyber) espionage, fraud, corruption and all forms of organised crime, but also territorial conflicts and blockades that threaten the Netherlands' strategic position as a transit hub and the security of supply of raw materials and energy. The International Security Strategy refers to climate security as a 'new' but increasingly topical theme.[47]

Although most of these themes are recognised as threats or risks, they have not been substantively addressed in specific domains of security or defence policy such as the raw materials policy or energy and climate policy. There are no references in defence policy to tasks or targets for deployability in relation to economic security, with the exception of cyber security (see Sect. 3.3.2 below).

Nevertheless, an indication has been given of the course to be taken. In *Houvast in een onzekere wereld. Lijnen van ontwikkeling in het meerjarig perspectief voor een duurzaam gerede en snel inzetbare krijgsmacht* [Maintaining a grip in an uncertain world. Lines of development in the multi-year perspective for sustainably ready and rapidly deployable armed forces] (14 January 2017), various aspects of flow security are discussed under the heading 'secure connections' (2017: 17) 'The Netherlands is a global hub', the report states, a 'system country' for which 'connectedness' and the 'hub function' are of the utmost importance. Protecting this vital infrastructure is crucial for preventing serious economic damage, threats to physical safety and social stability being undermined.

This hub function, whether it is 'gas roundabouts', the internet infrastructure or the Dutch ideal of economic security based on free trade and transparency, has already been challenged for some time by external developments. Russia and China, for example, have developed Anti Access/Area Denial (A2/AD) capacities in key regions. Russia has invested heavily in A2/AD capacities in areas extending from the North Pole to Syria, with concentrations in Kaliningrad and around the Crimea. China has created 'anti-access environments' in the seas of East Asia. These capacities (which include air defence, maritime activities, short- and medium-range ballistic missiles and cruise missiles) could not only effectively deny access to certain regions for an enemy's armed forces, but also its merchant fleet.

The takeovers of a number of large Dutch companies by state-owned companies and so-called sovereign wealth funds from Asia and the Gulf region in 2009 and 2010 sparked a debate about the tension between economic openness and economic security. There was an even greater shock when, in September 2013, América Móvil – which owned more than 30% of the shares in telecom company KPN – attempted to buy the remaining 70%. KPN owned a substantial portion of the fixed communication networks in the Netherlands, on which, in the absence of substitutes, vital government services and the networks of other telecom providers relied entirely, at least in the short term. If KPN were to fall into its hands, América Móvil would be able to exert pressure on the Dutch government, for example by threatening to shut down telecom services. It could also endanger the confidentiality of electronic data traffic.[48] To what extent could and should the Netherlands arm itself against such acquisitions and direct interventions in the interests of national security? Should it create a special framework for assessing investments or institute an investment review as Canada and the US have done? Similar discussions are being conducted in South European countries, where Chinese companies own large positions (but not majority shareholdings) in the electricity networks.[49]

In 2010, the Dutch General Intelligence and Security Service (AIVD) and the Ministry of the Interior and Kingdom Relations observed that there was little the Dutch government could do, even if such acquisitions were potentially harmful to national security interests. The AIVD pointed out that the Netherlands faced increased risks to national security because of its relative vulnerability to economic and technological espionage. Vital sectors of the infrastructure, such as telecommunication, energy and the aerospace industry, are increasingly exposed to spying activities by foreign intelligence services with an interest in gathering sensitive (commercial) know-how and information.[50] However, the ministries of Economic Affairs and Finance and representatives of the business community countered with the argument that foreign direct investment makes an unparalleled contribution to the Netherlands' economic security in the form of growth, prosperity and jobs. According to them, existing legislation was more than adequate in those rare instances where foreign investments could harm national security interests.[51]

Nevertheless, the case of KPN led directly to economic security climbing higher on the policy agenda. Two months later, an interdepartmental Economic Security

Working Group was established, chaired by the National Coordinator for Security and Counterterrorism (NCTV) and including representatives of the employers' organisation VNO-NCW. In April 2014, the working group produced a report entitled *Tussen naïviteit en paranoia* [Between naivety and paranoia], which contained a number of specific recommendations for safeguarding national security in the event of foreign takeovers and investments in vital sectors. The report stressed the undiminished importance of foreign acquisitions and investments for the Dutch economy, but was at the same time more insistent than earlier policy documents about the importance of weighing up general security interests in the event of foreign investments and takeovers. Assessments of that nature are 'unusual' in the Netherlands and 'take place on an ad-hoc basis' without 'ex-ante analysis to identify the risks to national security', the report said. In addition to ex-ante analyses, the authors called on the government to delegate ownership of the issue so that warnings from public and private security partners could be addressed and for a structural alliance of economic and security partners. The working group also said there was a need for expertise in the domain of economic security throughout central government.

The Minister of Justice and Security was adopting the recommendations to commission ex-ante analyses in two or three 'vital sectors' and to promote public-private cooperation, he said in a letter to the House of Representatives in June 2014.[52] But the minister also called for delineation of the theme of economic security. In addition to foreign takeovers and investments in vital sectors, the other 'priority topics' at that time were access to raw materials, protection of trade routes and (digital) espionage. At the same time, the Minister of Economic Affairs promised the House of Representatives that he would formulate additional powers, on the basis of advice from the ministers of Justice and Security, the Interior and Kingdom Relations and Defence, to allow him to evaluate the national security implications of changes of control of vital telecommunications infrastructure.[53]

In the *Voortgangsbrief nationale veiligheid* [Progress report on national security] in May 2015, however, the main emphasis was on formulating a uniform definition of vital infrastructure and identifying what constituted the Netherlands' vital infrastructure, partly with a view to promoting closer cooperation with the security regions and the business sector. Despite the growing attention to transnational cascade effects and external threats to economic security from espionage or acquisitions in vital economic sectors, there was still no question of economic security being firmly anchored in Dutch security policy as the Minister of Defence had hinted it would be in her reaction to the report *No Blood for Oil?* (2014) by The Hague Centre for Strategic Studies (HCSS).[54] At the beginning of 2017, a draft of a bill to prevent undesirable changes of control in the telecommunications sector was published for consultation. Under the proposed law, control over a telecommunication company may not fall into the hands of a party that is not or not exclusively acting on the basis of legitimate commercial interests, but (also) acting on the basis of geopolitical or criminal motives.

3.3.1 Energy and Climate Security

The policy letter *International Security. Turbulent Times in Unstable Surroundings* (14 November 2014) bears the traces of the Ukraine crisis. In it the government underlined the interconnectedness of security and the economy and referred to the implications of that crisis for energy policy. According to the government, the Russian intervention in Ukraine was prompted in part by the approaches that country was making towards the EU and which were obstructing Russia's plans for a Euro-Asian Union. Moreover, the annexation of the Crimea was accompanied by 'hybrid warfare', involving the use of unannounced, large-scale military exercises and rapid troop movements and secret support of separatist groups, as well as economic pressure. Since then, security of energy supply in Europe has been seen as an 'acute issue' that compels a review of the energy relationship with Russia.

The connection between Dutch energy policy and the international security situation was only made to a limited extent. In the policy letter, the government did not mention any specific steps the Netherlands would take to address what it had described as an urgent issue. It appeared to be leaving it to the EU to take the lead, although the EU had made little progress on the energy question up to that time because of the conflicting interests of the member states. The AIV's advisory report *De EU-gasafhankelijkheid van Rusland* [The EU's dependence on Russian gas] (June 2014) can therefore also be read as an appeal for a more active Dutch approach. The AIV observes in the report that geopolitical and security considerations receive far too little attention in European policy, but also in Dutch policy. The Netherlands allows its own trade and investment interests to prevail, so that too little consideration is given to reducing the dependence on Russia for energy. 'Energy policy in the Netherlands is determined primarily by economics', according to the AIV. 'Primary responsibility for it lies with the Minister of Economic Affairs. To a large extent, however, energy policy is also a matter of foreign policy. The Minister of Foreign Affairs should have a prominent voice in strategic investment decisions.'.[55]

This applies to an even greater extent now that the shale gas revolution in the US is not only having a direct impact on the competitive position of energy-intensive industries in Europe, but also indirectly has major long-term geopolitical implications. As analyses by organisations like the Netherlands Organisation for Applied Scientific Research (TNO) and the HCSS have shown, in time the diminishing American dependence on oil from the Middle East will translate into lower oil prices and a further shift of attention in American security policy from the Middle East and Europe to Asia and the Pacific region. As a result, the Pax Americana could also become a thing of the past without any new stabilisers appearing to replace it. Add to this the potentially disruptive effects of falling oil prices on economic and social stability in the southern EU member states and Russia and it is clear that a strategic energy policy for the Netherlands and Europe must take greater account of Dutch and European security policy and the European Neighbourhood Policy.[56]

It is not clear from the policy documents sent to the House of Representatives whether Prime Minister Mark Rutte's previous government was fully aware of the potential consequences of such shifts. In its reaction to the AIV report on the dependence on Russian gas, the government referred mainly to Europe. It said it supported the AIV's proposal to give the EU's High Representative for Foreign Affairs and Security Policy and the European External Action Service (EEAS) a greater say in energy policy. The government also promised to consider energy policy more explicitly in the European Neighbourhood Policy and to give the Minister of Foreign Affairs a voice in national decisions on strategic investments (Government Reaction of 7 October 2014). However, it remains to be seen what effect these statements will have in practice.

The International Security Strategy describes climate change as a driver of existing tensions and conflicts and goes on to advocate the introduction of early warning systems for problems relating to climate, water, food and other (scarce) resources.

The climate issue has been significantly higher on the Dutch security agenda than the energy issue, and for far longer. At international level, it has been the subject of widespread attention in the context of security policy since 2007 when it was raised in rapid succession in the UN Security Council, in *National Security and the Threat of Climate Change* (2007) by the American CNA Corporation, in the International Institute for Security Studies' *Strategic Survey 2007* and by the German *Wissenschaftlicher Beirat für Globale Umweltveränderungen*. The tenor of all of these reports was that the effects of climate change are manifested primarily in a decline in sources of fresh water and food production and an increase in extreme weather events, problems that threaten international stability and security because they exacerbate existing problems and conflicts. They could, for example, lead to a rise in the number of fragile states, cause economic growth to stagnate in certain regions, intensify the competition for scarcer resources and lead to growth in the number of (environmental) refugees and migrants.[57] Since then, climate change has also appeared in EU and NATO documents as a 'multiplicator' of existing international threats.[58]

The National Security Strategy in 2007 referred to the rising sea level as a security risk. The document spoke of 'ecological security'. It also referred to extreme weather conditions such as drought, heat and flooding as potential risks to national security. Climate change, and above all the rising sea level, were described as global issues that demanded a global response and were therefore not discussed any further.[59] In 2013, however, the National Risk Assessment included a scenario for flooding caused by the bursting of dikes tailored to the situation in the Netherlands.[60] The International Security Strategy in 2013 described climate change as part of a new 'complex security situation'. The government referred to the relationship between climate change, the 'conflicts over water, food, energy and raw materials', growing instability and an increase in the number of migrants. Conflict zones and fragile states are 'a breeding ground for terrorism, extremism and cross-border organised crime', particularly in the Horn of Africa, the Sahel region, North Africa and the Middle East. According to the International Security Strategy, 'preventive action' is essential 'in the face of new challenges resulting from climate change'. The urgency of the climate problems was further underlined in the policy letter *Turbulente Tijden in een Instabiele Omgeving* [Turbulent Times in Unstable

Surroundings], as was the importance of global action, legally binding agreements and global early warning systems. Climate change is expected to lead to growing demands on the armed forces to undertake missions in a steadily expanding range of circumstances and roles. The ministries of Defence and Foreign Affairs realise that this will also require internal changes in terms of infrastructure, facilities, training and testing environments and materiel. However, the climate perspective has still not been fully embedded in international security policy. The vast majority of steps that have taken up to now have been in the policy on development cooperation.[61]

In addition to pushing for integration of climate change in the formulation of national policy, the Netherlands recently took the initiative to establish an international forum to address issues at the intersection of climate change and security. The first international conference on Planetary Security (Peace and Cooperation in Times of Climate Change and Global Environmental Challenges) was organised at the Peace Palace in The Hague on 2 and 3 November 2015. A second conference followed on 5 and 6 December 2016. The Netherlands will act as facilitator of the new forum as the initiative is expanded and deepened in the coming years.[62]

3.3.2 *Security of Supply of Raw Materials*

Like energy and climate, raw materials have become part of the extended concept of security in the Netherlands. Figure 3.2 shows the growing mutual dependency between Russia and the EU in terms of raw materials. Dutch policy on raw materials was formulated against the background of the growing realisation that guaranteeing the security of supply of raw materials also represents an economic security interest

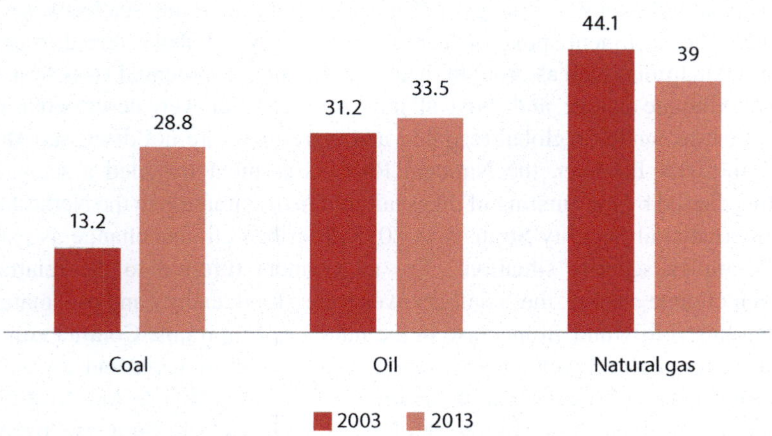

Fig. 3.2 Russia's share of EU imports of coal, crude oil and natural gas, as percentages of total European imports (Source: Clingendael, Strategic Monitor 2016, *Grootmachten en mondiale stabiliteit* [Great powers and global stability], 2016: 25)

for the Netherlands.[63] The Netherlands is an important transit country for commodities, has a large agro sector that depends on imported biotic raw materials, and a small but technologically advanced industrial sector which relies on flows of imported abiotic raw materials. A large number of multinationals with ties to commodity producers are also based in the Netherlands, and the country has a dense network of trade and diplomatic missions with which it occupies a powerful position in the force field of commodity flows.[64] The Dutch 'top sectors' of logistics, agrofood and chemicals in particular can profit from Europe's scarcity of raw materials thanks to their innovative capacity.[65] In 2011, the government of the time drew up a Policy Document on Raw Materials [*Grondstoffennotitie*] setting out an 'integrated strategy' for reducing the vulnerability of the Dutch economy.

Dutch policy represents a balancing act. On the one hand, it conveys the familiar aim of achieving the maximum possible in terms of open markets and free enterprise. On the other, it refers to the challenges posed by the climate and a geopolitical reality of protectionist state capitalism and market strength aimed at guaranteeing access to knowledge, technology and supplies of energy and raw materials. Security of supply is regarded as primarily the responsibility of the business community. The government can 'facilitate, stimulate, establish frameworks and coordinate'. Where 'the market does not work properly' the Netherlands must intervene via the WTO and the EU as far as possible.[66] The credo is 'European where possible, national where it creates opportunities".[67] At the same time, the pursuit of sustainability is an explicit prerequisite for security of supply in the longer term. The use of raw materials must not be accompanied by violations of human rights or damage to the environment.[68]

According to the progress reports to the House of Representatives, the government has facilitated various small-scale initiatives since 2011. A Special Envoy for Natural Resources has been appointed to actively promote cooperation with suppliers of vital raw materials. Platforms have been established to raise awareness of the problems and to promote knowledge sharing and technological cooperation, and Green Deals have been concluded with the business community with a view to removing obstructive laws and regulations.[69] However, researchers at institutes including the HCSS, TNO, the Netherlands Environmental Assessment Agency, the Council for the Environment and Infrastructure and the Rathenau Institute have found that there is still no integrated strategy for raw materials. In the first place, that requires a long-term perspective that provides the business community and research institutes with the certainty they need to make major investments. Secondly, it calls for more substantive coordination with the broader foreign and security policy, including economic diplomacy and development cooperation. And finally, the government needs to actively encourage companies and other actors in society to generate green growth and create a circular economy.[70]

At the beginning of 2017, the National Raw Materials Agreement was signed by 180 parties. They are now preparing plans to accelerate the transition to a fully circular economy by 2050. The transition from a linear economy (in which raw materials are ultimately converted into waste) to a circular economy (in which waste is re-used as a raw material) will have a positive effect on global public goods such as the climate and biodiversity. It will also yield competitive advantages and is therefore also important for the country's economic position. However, the policy on raw materials is still not adequately geared to the context of an extended security policy. Geopolitics and this country's security are also at stake here via the (negative) impact on commodity-rich countries, including developing countries.[71]

3.3.3 Cyber Security

As in other countries in Europe, the expansion of the Dutch security agenda to the cyber domain is now well advanced. Furthermore, cyber security encompasses all of the 'objects' of security: the state, the relevant actors in society and the individual.[72] All of the vital interests that the Dutch government should protect (territorial, physical, economic and ecological security, political and social stability, and the international legal order) are so deeply permeated by digital systems that the digital domain itself can be regarded as a vital interest.[73] In 2011, the NCTV drafted the first National Cyber Security Strategy (NCSS1) for the government. The strategy referred to the international sources of insecurity, but like the National Security Strategy, it was primarily concerned with preventing disruption in the Netherlands itself. The government explicitly sought to achieve that through cooperation with companies, institutions and citizens, since precaution is a shared responsibility. The second National Cyber Security Strategy (NCSS2) in 2013 bore all the hallmarks of a fully-fledged strategy. It explicitly made the connection with the international domain. Taking inspiration from the integrated 3D approach, the authors said that capacity-building was needed to guarantee cyber security against both national and international threats. In an accompanying Action Programme 2014–2016, the strategic orientation was translated into specific objectives and actions to be taken by the relevant actors. The Netherlands wishes to be in the vanguard of international efforts to combat cyber crime, for example through further harmonisation of legislation relating to the investigation of that form of crime and by strengthening Europol's European Cybercrime Centre.[74] The Netherlands also played an important role in the establishment of the Global Commission on the Stability of Cyberspace (2017). In 2012, an integrated Defence Cyber Strategy was drawn up focusing on internal and external cyber security. The Defence Cyber Command was established in 2014 to provide the necessary capabilities for the defence of digital systems.[75]

The commander of the Defence Cyber Command, Hans Folmer, has expressed concerns about the existence of advanced capabilities to manipulate information, for example with real-time editing of video images in order to mislead citizens, military personnel and politicians. The 80 person-strong Cyber Command that he leads prepares for defence in the event of various scenarios, such as a cyber attack on the energy supply, payment systems or flood barriers. Since recently the Command has also been developing its own offensive capabilities designed to eliminate hostile communication systems.[76]

The Netherlands' 'digital delta' is highly developed, but it is vulnerable to cyber attacks. Cyberspace is increasingly the setting for geopolitical and terrorist threats and the development of the Internet of Things is greatly increasing the number of potential targets.[77]

3.4 Integrated Approach Needed More Than Ever

The expansion of the security agenda and the greater interconnectedness of internal and external security in an interdependent network society are having an evident impact on Dutch security policy. Two developments stand out. The first is the focus on stabilising fragile states and the integrated approach to security and development during missions outside Dutch territory. The second is the emergence of flow security, or economic security, as an area of concern in strategic documents and instruments. Both developments contribute to the expansion of the role of the armed forces as a structural security partner for civil actors.

The main points of the policy on fragile states have not changed since 2008, despite various indications that the approach taken has produced very limited results up to now. What is lacking is a detailed analysis of the underlying causes and an approach geared to prevention and the long term.

Economic security and flow security have only appeared higher on the security agenda fairly recently and are consequently not yet truly embedded in government policy. This is undoubtedly due to the Netherlands' free-trade orientation and traditional aversion to geopolitics. The attention to (cyber) espionage and foreign takeovers of companies that could be of strategic importance, for example, really only came to the forefront of political attention with the América Móvil affair, despite earlier warnings from the AIVD and NCTV. The policies on energy, climate and raw materials have also only received serious attention in discussions of security policy in the last few years. On the other hand, the Netherlands, partly due to its own high internet density and vulnerability to cyber attacks, has for some time had a comprehensive, future-proof cyber strategy that embraces both the national and international domain, as well as a Defence Cyber Strategy and a Defence Cyber Command specifically for the defence forces.

With the recent deterioration in the international security situation, an integrated approach to security – with a defence policy derived from it as one of the instruments – is needed more than ever. In the immediate vicinity of Europe, the 'new reality' of a ring of instability[78] calls for a policy that goes beyond a re-evaluation of (collective) defence and repressive treatment of symptoms. The integrated use of a variety of instruments (diplomacy, development cooperation, strategic economic policy, defence, intelligence, etc.) is also necessary to structurally address the underlying causes of instability and insecurity along the eastern and southern borders of Europe. In the process, it is essential to take account of equally integrated or hybrid threats from actors that use both conventional and non-conventional, both military and paramilitary and civil, and both open and secret means, to strike at vulnerabilities and to serve their strategic interests.

Endnotes

1 Castells, M. (1996); Goede, P.J.M. de et al. (2001).

2 The Uppsala Conflict Data Program makes a distinction between 'state-based' and 'non-state' conflicts. See Uppsala Conflict Data Program: http://ucdp.uu.se/

3 Pilbeam, B. (2015a, b).

4 WRR (2001: 164–5).

5 Parliamentary Documents II 30,075, no. 1, 2004–2005, 22 March 2005.

6 Korteweg, A.R. (2011, 284–290); Klep, C. (2012); Grandia, M. (2015, 115).

7 Grandia, M. (2015, 119–120).

8 Between August 2004 and January 2005 only informal talks were conducted between the small group of Dutch troops and their British and Canadian colleagues. It was the end of January before the Military Operations Steering Group was informed of these talks. Grandia notes that General Jim Jones, NATO's Supreme Allied Commander Europe (SACEUR) was surprised by the dynamic of the bilateral talks: "They hammered out the whole thing without NATO assistance, behind closed doors…We were not aware of the details" (Grandia 2015: 116). The head of military affairs and the head of diplomacy at the Dutch mission to NATO also had scarcely any involvement in the initial planning of the mission. The House of Representatives was informed about the plans for the mission in June 2005.

9 Grandia, M. (2015: 121).

10 Grandia, M. (2015: 139–150); Policy and Operations Evaluation Department of the Ministry of Foreign Affairs, IOB (2013).

11 Amersfoort, H. (2016: 219).

12 Dimitriu and De Graaf observe that the concept of counterinsurgency – with the accompanying offensive operations – figures in all ISAF documents and also played a role in the parliamentary debates in France and Britain. The Dutch government avoided the use of this term (Dimitriu & De Graaf 2011).

13 Brocades Zaalberg, T.W. (2008: 121).

14 IOB. (2013: 109); Korteweg, A.R. (2011: 292–6); AIV. (2009).

15 IOB. (2013: 94).

16 Lijn, J. van der (2011: 32–35).

17 IOB (2013: 111–115).

18 Gruiters, J. (2011: 3); IOB (2013: 112); Dimitriu, G. & Graaf, B. de (2011).

19 Gruiters, J. (2011: 4); IOB (2013: 113).

20 Gruiters, J. (2011: 6); IOB (2013: 93–95).

21 Gruiters, J. (2011: 6).

22 IOB. (2013: 95).

23 Lijn, J. van der (2011: 34–44).

24 Parliamentary Documents II 2008–2009, 31,787, no. 1, 7 November 2008.

25 AIV. (2009).

26 AIV. (2009: 53).

27 Parliamentary Documents II 2011–2012, 32,605, no. 94, May 2012.

28 For an evaluation of progress with the New Deal, see International Dialogue on Peace-building and State-building (2014) and Van Veen and Dudouet (2017). On 5 April 2016, the parties to the dialogue renewed their commitment to the principles of the New Deal in the *Stockholm Declaration on Addressing Fragility and Building Peace in a Changing World.*

29 Multi-year Strategic Plans 2014–2017: https://www.rijksoverheid.nl/documenten/rapporten/2014/02/05/meerjarige-strategische-plannen-mjsp-2014-2017

30 The focus of each organisation is as follows: (1) Interpeace: operates in complex conflict zones (early peace-building) where the UN has less expertise and/or access and is able to link peace-building at local level (track 3) and at civil-society level (track 2) with peace negotiations at political level (track 1); (2) International Crisis Group: an international watchdog operating in many (potential) conflict zones in almost 100 countries in all; (3) International Alert: is engaged in developing and influencing policy and capacity building in relation to peace-building. A leading actor in the area of gender in conflict, conflict analyses and conflict sensitivity; (4) Centre for Humanitarian Dialogue: a leading actor in track 1 and track 2 mediation. Its most important activities involve strictly confidential mediation; (5) International Centre for Transitional Justice: sets the international standard for the development and application of Transitional Justice. The Ministry of Foreign Affairs has not put these activities out to public tender and does not ask the partners to concentrate explicitly on one or more conflicts, but does hold talks with each of the five organisations every year. The Ministry of Foreign Affairs also commissions research by the Clingendael Conflict Research Unit, mainly for the purpose of analysis and formulation of policy options. It also collaborates with EU delegations in the context of EU Joint Programming. Joint programming between the EU and member states encompasses (1) joint analysis, (2) joint identification of priorities and (3) allocation of work.

31 Parliamentary Documents II 2013–2014, 22,112, no. 1778, 17 January 2014.

32 Parliamentary Documents II 2014–2015, 33,694, no. 7, 17 November 2014.

33 Letters to the House of Representatives regularly contained references to the 'successful 3D approach' (Development, Diplomacy and Defence) which deserved to be continued: "The building and strengthening of the legal order in developing countries is crucial for safeguarding our interests. The Netherlands has, via an integrated and innovative approach (think of the 3D approach in Afghanistan), achieved results in the past that can be built on" (Parliamentary Documents II 2010–2011, 32,500 V, no. 15). The 3D or integrated approach is not a Dutch invention, despite repeated references to the 'Dutch approach' and the Netherlands' pioneering role. The British version was originally developed

by the Ministry of Defence on the basis of experiences in Bosnia, Kosovo and Sierra Leone, where development-oriented, humanitarian and political activities were combined. The ministry formulated four basic principles:

- proactive involvement prior to the crisis in order to properly assess the situation on the spot, to be able to interpret early warning signs and to start planning as soon as possible in order to have a timely presence;
- a shared understanding of the situation in order to operate as effectively as possible;
- thinking in terms of results and what is needed to achieve them. The planning and activities must therefore be concentrated on a single goal and success must be measured against shared benchmarks of effectiveness;
- joint action: familiarity with each other's institutions and networks and personal trust, in addition to integrated information management, help enhance the sharing of information and the joint performance of activities (House of Commons 2010).

In order to strengthen interdepartmental operations, in 2004 the British government introduced a budget for conflict prevention, stabilisation and peacekeeping jointly controlled by the Ministry of Defence, the Foreign and Commonwealth Office and the Department for International Development. A joint Post-Conflict Reconstruction Unit (since renamed the Stabilisation Unit) was also created to facilitate the drafting of joint analyses and plans. This unit was also given responsibility for establishing a so-called Civilian Stabilisation Group, a pool of more than a 1000 citizens and police officers from which at least 200 people could be deployed for missions at any one time. The integrated approach is also included in the British Defence Doctrine and the first National Security Strategy (House of Commons 2009).

34 House of Commons (2010a, b).

35 Kasselmann, H.J. (2012); Kitzen, M. (2008).

36 The term 'counterinsurgency' sounds militaristic, but, according to the US Government Counterinsurgency Guide, strictly speaking it involves a 'blend of comprehensive civilian and military efforts designed to simultaneously contain insurgency and address its root causes'. Cited in B. de Graaf (2010) The Dutch mission in Uruzgan. Political and military lessons, *Atlantisch Perspectief* 2010 (34), no. 7.

37 Lijn, J. van der (2011: 27–28; 50–56); Matthijsen, C.J. (2014); Homan, C. (2014: 57).

38 *Leidraad Geïntegreerde Benadering* (2014: 11–14).

39 Wittkowsky, A. & Wittkampf, U. (2013: 2).

40 As De Spiegeleire puts it: "It then immediately becomes the subject of a number of political bargains between these players. At the same time, this domestic bargaining process is also thrown in a number of multilateral consultations where similar political horse-trading takes place between different countries.

The political games that ensue in the best case yield a political 'decision' that specifies a set of fairly vague political objectives. In case a military contribution is requested, the military is tasked to come up with an operational plan based on this political guidance." (Spiegeleire 2014: 47).

41 Lijn, J. van der (2011: 60).

42 Coning, C. de & Friis, K. (2011); IOB. (2013); AIV. (2009).

43 Spiegeleire, S. de et al. (2014: 1).

44 Esch, J. van, Jong, S. de & Ridder, M. de (2014: 29).

45 Esch, J. van, Jong, S. de & Ridder, M. de (2014).

46 NSS. (2007: 10).

47 ISS. (2013: 10).

48 Parliamentary Documents II 2013–2014, 24,095, no. 368, 10 June 2014.

49 *Infrastructuren als wegbereiders van duurzaamheid*, WRR Working Paper no. 12.

50 Parliamentary Documents II 2009–2010, 30,821, no. 11, 2 April 2010; Esch, J. van, Jong, S. de & Ridder, M. de (2014: 33–360.

51 Esch, J. van, Jong, S. de & Ridder, M. de (2014: 35).

52 Parliamentary Documents II 2013–2014, 30,821, no. 22, 10 June 2014.

53 Parliamentary Documents II 2013–2014, 24,095, no. 368, 10 June 2014.

54 Parliamentary Documents II 2013–2014, 33,763, no. 44, 28 May 2014.

55 AIV. (2014: 19).

56 HCSS & TNO (2014b).

57 Homan, C. (2005).

58 Schaik, L. van et al. (2015: 34).

59 National Security Strategy (2007).

60 National Risk Assessment (2013).

61 The Dutch policy on development cooperation links the problem of climate change to two of the four so-called spearheads, namely water and food. In practice, this means that attention will be devoted to the implications of potential climate risks for food and water projects in each of the partner countries named in the multi-year security plans. There is also a review of the opportunities for investment in future projects that will contribute to strengthening the resilience of the countries against climate change.

62 Ministry of Foreign Affairs (2016).

63 Parliamentary Documents II 2010–2011, 32,852, no. 1, 15 July 2011.

64 HCSS. (2011: 12).

65 Weterings, R. et al. (2013: 65).

66 Parliamentary Documents II 2010–2011, 32,852, no. 1, 15 July 2011.

67 Parliamentary Documents II 2010–2011, 32,852, no. 1, 15 July 2011.

68 Krom, A. et al. (2015: 91).

69 Parliamentary Documents II 2012–2013, 32,851, no. 15, 4 July 2013; Krom, A. et al. (2015: 91).

70 HCSS. (2011); Weterings, R. et al. (2013); Krom, A. et al. (2015); Buijze, G. (2013); Netherlands Environmental Assessment Agency (2014).

71 Ridder, M. de (2017).

72 Recent research carried out for the HCSS on the basis of recent indices showed that the Netherlands, together with the UK and the US, are regarded as being best prepared for threats to cyber security (Gehem et al. 2015: 71–72).

73 Ducheine, P. (2016).

74 Ministry of Justice and Security (2013).

75 Parliamentary Documents II 2014–2015, 33,321, no. 5, 23 February 2015.

76 '"Cybergeneraal" bezorgd over videomanipulatie' ["Cyber general" concerned about video manipulation], *NRC* 28 February 2017.

77 National Cyber Security Centre (2016); Rathenau Institute (2017).

78 AIV. (2015, 2016).

Chapter 4
The Netherlands and the Extended Concept of Security: The Rise of Security Strategies

Ernst Hirsch Ballin, Huub Dijstelbloem, and Peter de Goede

4.1 Introduction

With the extension of the concept of security that has occurred over the last few decades, formulating security policy is inherently problematic. Even if one takes a narrow state-centric view of security as military security behind a territorial line of defence, threats are already enormously difficult to assess. After all, states do not readily provide information about their military capacity and gauging their intentions is a problem in itself. Moreover, those intentions are not static, but can change over time. States can also mislead one another with regard to their capabilities and intentions. The history of international relations is replete with examples of threats being underestimated, exaggerated or incorrectly interpreted and of political failure to respond to accurate estimates and pinpoint strategic analyses. Furthermore, states also have to estimate how other states will react to their own actions and omissions. History is therefore also filled with examples of unintended consequences.[1]

However, the complexity and dynamic of security policy increases exponentially when national security policy has to be coordinated with the policies of allies, for example at EU and NATO level. When internal and external security are interconnected by numerous transnational relationships. When, in addition to military security, human security and flow security also have to be taken into consideration. And when security policy can only be geared to a limited extent to specific, known threats and security also has a subjective, socially constructed dimension.

Wherever one looks, strategy formulation is a tried and trusted response in attempting to get a grip on this extremely complex and dynamic security environment. This chapter opens with a general outline of the rise of national security strategies and the related approaches and instruments (Sect. 4.2).

E. Hirsch Ballin (✉) · H. Dijstelbloem · P. de Goede
WRR, Scientific Council for Government Policy, The Hague, Zuid-Holland, The Netherlands
e-mail: ballin@tilburguniversity.edu; h.o.dijstelbloem@uva.nl; peter.goede@minbzk.nl

E. Hirsch Ballin et al. (eds.), *Security in an Interconnected World*,
Research for Policy, https://doi.org/10.1007/978-3-030-37606-2_4

We then discuss the steps taken by the Netherlands in relation to formulating a strategy. Since 2007, the Dutch government has been more systematic in its endeavours to explore the internal and external security environment. On further reflection, however, it can be seen that the Netherlands does not make full use of the strategic instruments and the underlying philosophy (Sect. 4.3), which leads to a number of conclusions with regard to how policy-making in relation to security can be improved (Sect. 4.4).

4.2 Strategic Instruments: Getting a Grip in a Complex and Dynamic Security Environment

4.2.1 The Rise of National Security Strategies

Strategy formulation has a long history. China's Sun Tzu wrote about the art of war as early as the sixth century BC, starting a tradition in which he has been followed by writers such as De Jomini, Machiavelli and Von Clausewitz.

The US has had a national security strategy since the 1950s, when President Eisenhower established a Planning Board within the National Security Council. Elsewhere, national security strategies only really took off at the beginning of the twenty-first century, often in response to the changing security environment and the growing interconnectedness of internal and external security (Box 4.1).

Box 4.1: The Rise of National Security Strategies

Lithuania (2002).

Poland (2003).

Canada (2004) *Securing an Open Society: Canada's National Security Policy.*

Finland (2004).

Slovakia (2005).

France (2008) *Défense et Sécurité nationale. Le Livre blanc.*

Australia (2008) *National Security Statement.*

United Kingdom (2008) *The National Security Strategy of the United Kingdom.*

Spain (2011).

Hungary (2011).

Belgium, Denmark, Germany, Italy and Sweden are among the countries that have not drafted a security strategy. In Germany, there is a discussion underway at the moment about a 'White Book' for the Ministry of Defence (*Bundesministerium der Verteidigung* 2016) and whether to assume a greater role in international crisis management, in the context, among other things, of a project by the Ministry of Foreign Affairs, *Aussenpolitik Weiter Denken*, and in response to the report of the Rühe Commission (whose full title is *Kommission zur Überprüfung und Sicherung der Parlamentsrechte bei der Mandatierung von Auslandseinsätzen der Bundeswehr*, June 2015).

There are important similarities between the various national security strategies mentioned above:

- *The connection between internal and external security is stressed.* Transnational issues and interdependencies have blurred the traditional distinction between internal and external security.
- *The object of security is not just the state.* National security concerns not only the state and its vital institutions, but also society and individual citizens.
- *Security is more broadly defined.* National security has many manifestations, including economic security, energy security, maritime security and cyber security. The threats and risks to security are equally polymorphous.
- *The need for a whole-of-government approach is acknowledged.* Some strategies (including those of the US, Australia, Slovakia, Spain and the United Kingdom) have introduced a so-called whole-of-government approach.[2] They serve as a blueprint for a variety of actors in the security domain, not just the armed forces. Security is therefore no longer determined entirely by military capacity, but also by social resilience (the capacity to deal as effectively as possible with threats).
- *The strategies are public documents, primarily targeted at politicians and the public, in the country itself and abroad.* The strategies analyse the security environment and address values, interests, objectives and means.

There are also important differences, some of which are connected with historical, geographic and cultural factors. There are country-specific threats and risks, such as demographic trends or economic instability; country-specific objectives, such as the protection of a country's own national minorities beyond the national borders; and country-specific priorities, such as a focus on the neighbouring region. The extent to which national security strategies set priorities, make choices and contain specific measures and guidelines also varies greatly. A national security strategy is ultimately a combination of all these factors.[3]

The proliferation of national security strategies was in part a reaction to the changes in the security environment, but also partly a recognition of the numerous possibilities and functions of strategy-formulation processes (see Box 4.2). Accordingly, the strategies vary as regards their form and content and cannot be seen in isolation from the specific context in which they were formulated.

Box 4.2: Functions of Strategy Formulation

– *Strategy formulation as a method of expressing the relationship between goals and resources.* If goals are carefully formulated and correctly combined with resources, there is a chance of success. As Colin Gray puts it: "Strategy is a functional necessity for every human society, since all political communities need a security that must entail endeavour to match political ends with good enough available means employed in tolerably effective ways".[4] Gray stresses the political context of strategy formulation. The goals set derive their significance from politics, just as the allocation of means is, by definition, a political question.

- *Strategy as a narrative*. Lawrence Freedman defines strategy as 'the art of creating power'.[5] Powers of persuasion are an essential aspect of strategy. Freedman therefore attaches great value to strategy as a narrative in a world in which waging war has become a choice ('wars of choice').[6] Alan Stolberg further elaborates on this. National security strategies can promote the consensus within the government, make it easier to secure parliamentary approval for the allocation of resources and serve as a strategic tool for communicating with the country's own population and with state and non-state actors beyond the national borders.[7]
- *Strategy as a formulation of realistic political goals*. The military interventions in Afghanistan and Iraq (in which the Netherlands was involved in various ways) did not proceed smoothly. Military objectives were achieved initially, but the unplanned follow-up – the creation of institutions and a state – encountered serious difficulties. Particularly in the United Kingdom and the US, the persistent problems in Afghanistan and Iraq led to calls for politicians to start taking strategic thinking seriously again. In the Netherlands, Isabelle Duyvesteyn expressed her criticism as follows: "We have forgotten how to formulate feasible political objectives, to attach realistic military plans to them and to apply the two of them in balance and proportion".[8] She referred to 'strategic illiteracy'.
- *Strategy as a 'way of coping'*. A strategy addresses a problem in a dynamic environment with opponents, but also with allies. Negotiation and compromise are therefore the rule. There is a desired end result, but in practice it is about proceeding to the next phase.[9] A strategy is not a timetable with a guaranteed time of arrival at the desired location, but a tool to help in holding your own in a dynamic environment and to influence it. Kramer draws the same conclusion with regard to irregular warfare. Conflicts in fragile states are 'wicked problems'. There is no consensus about the underlying causes or about the solution. Intervention provides no certainty about the outcome and has unforeseen consequences. This calls for imperfect strategies whose central objective is to be 'good enough'.[10]
- *Strategy as 'grand strategy'*. The changed security environment calls for a whole-of-government approach that allows for the use of a wide range of instruments to address a variety of threats and risks. There are similarities between national security strategies based on this principle and 'grand strategy' in its original sense – the deployment of the state's entire resources for the purpose of winning a war. Colin Gray observes a revival of 'grand strategy'.[11]
- *Strategy as a guideline for the armed forces*. Many armed forces are confronted with major problems because of the expansion of the security agenda and the proliferation of risks. In light of the uncertainty about the future, they insist on retaining the widest possible range of capabilities ('capability based defence planning'). This approach appears sensible, but imposes great demands on affordability. It is also problematic in strategic terms, because the balance shifts from goals to resources.[12] However, making

choices on the basis of quantified risks is also not a solution, according to Gray, simply because no method is capable of foretelling the future in detail.[13] Furthermore, such an approach wrongly disregards the political context. An overarching strategy with clear goals and means is therefore the only basis for sound defence planning.

- *Strategy as a process*. Linking goals to resources is a continuous process in which it is not only the outcome that matters. President Eisenhower's motto 'it's not about the plan, it's all about the planning' is therefore widely endorsed in the literature and in practice. According to the participants in the process, the great added value of the drafting of the American National Security Strategy was that the various actors in the security domain worked together in formulating visions and perspectives for action.

In a reaction to the interventions in Afghanistan and Iraq, the American Center for Strategic and Budgetary Assessments (CSBA) concluded that the strategic capacity of the American political and military elite had been in decay for decades. The primary reason, according to the CSBA, was a misapprehension of the nature of strategy. Strategy is not the same as compiling a list of desired goals. The point of a strategy is to identify how those goals can be achieved despite limited resources, bureaucratic resistance, political considerations and uncertainty about the actions of opponents and the effect of the chosen strategy. This demands competent strategists, sufficient time and attention from the political leadership and effective structures for formulating strategy.[14]

In the United Kingdom, a House of Commons committee published a powerful plea for strategy formulation, but also expressed criticism of the approach that had been adopted up to then. The committee formulated ten principles of good strategy making (see Box 4.3).

Box 4.3: Principles of Good Strategy Making (United Kingdom)

1. investment of time and energy by ministers to create an 'appetite' for strategic thinking;
2. definition of long-term national interests, both domestic and international;
3. consideration of all options and possibilities, including those which challenge established thinking and settled policies;
4. consideration of the constraints and limitations which apply to such options and possibilities;
5. a comprehensive understanding of the resources available;
6. good quality staff work to develop strategy;
7. access to the widest possible expertise beyond government;
8. a structure which ensures the process happens;
9. audit, evaluation and critical challenge; and
10. parliamentary oversight to ensure scrutiny and accountability.

Source: House of Commons, Public Administration Select Committee, Who does UK National Strategy?, 2010b: 13.

Bailes[15] observed that small countries in particular benefit from strategy formulation and anticipatory capacity because they find themselves, by definition, in a complex and demanding security environment and have more limited resources. The chance of formulating a successful strategy is greater if the following questions can be answered in the affirmative:

- Is there an explicit risk assessment process with high-quality information in place? Is an effort made to be objective or are priorities influenced by traditions, over-generalisation of recent experiences, social unease or excessive attention to shocking incidents? Are informed non-state actors involved? Who defines security issues and are they the correct actors?
- Is there a common view of security and defence? Are perceptions, interests and values united? After all, divisions can lead to contradictions or unstable compromises that undermine the influence of small countries. To be heard, one has to take a strong stance and be consistent.
- What are the official structures for assessment, decision-making and implementation with respect to security policy? Is there a clear decision-making hierarchy? Is there sufficient coordination between the various actors? Is there a nerve centre where decisions can be made? The assumption that formal structures are unnecessary and affairs can be managed decentrally is a typical weakness of small countries.[16]

4.2.2 Whole-of-Government Approaches

The rise of strategy formulation has been accompanied by the ascent of the whole-of-government approach to security. After all, there is a wide range of actors inside – and outside – the government operating in the socialised security domain.[17] Accordingly, there is a need for structured cooperation between all of the relevant actors, including NGOs, the business community and knowledge institutes (hence, this is also referred to as a whole-of-society approach).

The link was quickly made between the whole-of-government approach and the integrated approach to security and development in fragile states (see Chap. 3). But efforts were also made, in Canada and Singapore for example, to strengthen coherence and cooperation in relation to national security. While those countries still explicitly place the necessity of closer interdepartmental cooperation and coordination in the context of national security, Australia took the step of adopting an integrated whole-of-government approach to national and international security in its National Security Statement at the end of 2008. A number of countries, including the US and the United Kingdom, have since followed Australia's example.

In short, the whole-of-government approach has entered the security domain in the last decade, first in addressing problems in fragile states and later in the realm of internal security. This provided a further impulse for a comprehensive approach to internal and external security. Parliaments play an important role in promoting the whole-of-government approach, particularly during the phase when the strategic vision is being formulated (Box 4.4).[18]

Box 4.4: Lessons of Whole-of-Government Approaches
Five lessons were drawn from an evaluation of the approach adopted in Singapore:

1. The government should make use of guiding documents that outline the strategy and ensure coherence;
2. Leadership is crucial in choosing and fleshing out a whole-of-government approach;
3. Ownership of whole-of-government projects is crucial;
4. The capacity to work in cross-departmental teams cannot be taken for granted. It requires the selection and training of suitable individuals;
5. A whole-of-government culture cannot be taken for granted. Efforts must be made to build mutual trust and cooperation in networks.

Source: Singapore National Security Coordination Secretariat http://www.nscs.gov.sg/public/home.aspx

Research in a number of northern European countries, including the Netherlands, identifies the following additional requirements of a whole-of-government approach:

1. consensus on goals, resources and methods;
2. adaptation of procedures and structures;
3. an overarching, supra-departmental structure and thematic units to prevent compartmentalisation;
4. an overarching strategic vision to prevent ministries from following their own course on the basis of mandates;
5. a culture of cooperation and listening to one another.

Source: Jermalavicius, Pernik and Hurt (2014).

4.2.3 Foresight Studies and Risk Assessment: Looking Differently at Security

In addition to strategy formulation and a whole-of-government approach, foresight studies and risk assessments have also been introduced as tools for coping properly with the extended security agenda. These are (clusters of) strategic instruments designed to identify possible, but uncertain developments. Foresight and risk assessment both reflect a view of security dominated by uncertainty about the future and (potential) risks rather than traditional, known threats.

Foresight The interest in foresight in the security domain is connected with the increased uncertainty about the complex security environment, which makes it difficult to reach a consensus on the most important threats and risks and the allocation of scarce resources. Foresight is not the same as predicting the future, which is in any case impossible. Thinking in terms of prediction would imply that the future is already determined. That is not the case – the future is not fixed. The future is open, but is also not empty, because the present and the past cast their shadows over it.[19] The method of exploring various possible futures (in contrast to predicting a future that is free of surprises) is generally referred to as foresight, which is defined as "… the process of developing a range of views of possible ways in which the future could develop, and understanding these sufficiently well to be able to decide what decisions can be taken today to create the best possible tomorrow".[20] The approach was developed in the military domain in the US. The Rand Corporation, one of the world's first commercial think tanks, played a decisive role in its development, devising the Delphi method among others. Most other developed countries have possessed – usually separate – civil and military foresight capacities for policy development for some time.[21]

A number of countries now publish trend reports in which a series of 'new' security threats and risks are assessed. In the US, the National Intelligence Council publishes a Global Trends report after every presidential election. President Trump received the sixth edition in December 2016 (*Global Trends* 2035).

The first phase in a strategic foresight study involves gathering information by means of horizon scans and/or 'early warning' systems. These tools enable researchers to pick up signals so that strategic surprises can be avoided and measures can be taken in time. In the second phase, the assembled information is analysed and an outline is produced of the outcomes of possible developments. In the third phase, options are fleshed out on the basis of scenarios, whereupon actions can be undertaken, some in the form of precautionary measures and some in the form of measures designed to bring about a desired scenario. Experience has shown that the added value often lies in the process itself as the participants develop new networks and ideas and share their views. Critical success factors include a whole-of-government approach to ensure that all the available information is collected, the involvement of external expertise (think tanks, universities, businesses and civil society), a thorough and reliable process with findings that are respected, and the intellectual freedom and political scope to challenge conventional ideas.[22]

Assessing Risks The focus on adopting measures to contain or control risks that are regarded as unacceptable is not a product of what the sociologist Beck[23] called the modern 'risk society', but has a lengthy history in domains such as fire safety, water and food security, infectious diseases and hazardous substances.

However, risk management has really taken off since the 1990s,[24] with the business community leading the way. Governments followed, partly in response to incidents and the political and public response to them.[25] The growing attention to known, calculable risks (risk as a function of probability x impact) has in fact gradually resulted in an approach that addresses a steadily growing number of conceivable, but difficult to quantify, uncertainties.

The thinking from a risk assessment perspective has had a major impact on the security agendas of states and international organisations like the EU and NATO. For example, risks and their management dominated in the European Security Strategy in 2003 (*A Secure Europe in a Better World*), and they occupied a prominent position in NATO's Strategic Concept in 2010 (*Active Engagement, Modern Defence*), in addition to the classical threats for which the obligation of collective defence applies.[26]

Partly because of this process, an 'uneasy peace' – in the words of Edmunds[27] – has arisen since the 1990s, in which the logic of threat has been replaced by that of risk. Although Edmunds endorses Beck's analysis of the emergence of the risk society, he makes a direct connection between the absence of major threats to Western countries and the dominance of risks in Western thinking on security. In these countries insecurity is measured mainly by what *could* happen in a context of uncertainty and complexity. It is perhaps too soon to judge whether Edmunds is right. Threats have returned after an absence, in the form of Putin's Russia and the IS jihadists. Moreover, the broad security agenda encompasses numerous and varied security risks. Perhaps two logics will co-exist in the security domain: one based on threats and the other based on risks.[28]

4.3 The Netherlands and the Strategic Instruments

Like many other countries, the Netherlands has started using the aforementioned strategic instruments (strategy formulation, the whole-of-government approach, foresight and risk assessment) in response to the changing security environment. This section describes and assesses how the Netherlands uses the available instruments.

4.3.1 Strategy Formulation: Separate Strategy Documents for Internal and External Security

Up to now the Netherlands has published two separate security strategies: one for internal security (the National Security Strategy, 2007) and one for external security (the International Security Strategy, *A Secure Netherlands in a Secure World*) [Veilige Wereld, Veilig Nederland], 2013).

The National Security Strategy is based on a comprehensive definition of security[29] and places the Netherlands in an international context, but is *de facto* confined to the country's internal security. The themes covered in 2007 were climate change, energy security, ICT breakdowns, polarisation and extremism, criminal infiltration of mainstream society and serious accidents. The National Security Strategy is primarily an instrument for preventing social disruption in the Netherlands.

The International Security Strategy is explicitly confined to international – external – security, with a reference to the National Security Strategy for internal security. That is noteworthy, because the existence of two separate security documents is out of tune with the government's recognition that internal and external security are interconnected (as well as being irreconcilable with the integrated approach to security and development issues at EU, NATO and UN level) (Box 4.5).

Box 4.5
The war in Syria is a perfect illustration of the blurring of the boundaries between internal and external security. Distances in time and geography play scarcely any role in that war, and not only because of the possibility of jihadists and Da'esh fighters returning from Syria in the stream of asylum seekers. At the height of the fighting around the city of Kobani in the north of Syria, for example, there were also confrontations between Kurds, Turks and Syrians in the Schilderswijk district of The Hague. The local authorities had to respond to them with 'crisis diplomacy'. According to Ko Colijn, the institutional approach to security is still failing to keep pace with these facts: "The Ministry of Foreign Affairs is responsible for external security, with the Ministry of Defence as the executive body; the Ministry of Justice and Security is responsible for internal security. They do what they can, but there is still no integrated security policy: everyone makes their own policy documents and risk assessments".[30]

To this day the government has not seen any reason to end this situation of separate worlds. The policy letter *Turbulent Times in Unstable Surroundings* (dated 14 November 2014), which could be described as a sort of updating of the International Security Strategy, mentions closer interdepartmental cooperation at the intersection of internal and external security, but leaves it at that. Not a word was devoted to the subject in the letter to the House of Representatives on the further development of the National Security Strategy (dated 12 May 2015).

The choice to maintain the status quo is probably a pragmatic one. However, the consequence is that, in contrast to many other countries, the government, politicians and the public in the Netherlands do not have an integrated security document. Conversely, there is in fact such a document for the overseas parts of the Kingdom.

It is easy to guess the consequence of the existence of separate security documents. In April 2014, the Clingendael Institute published the findings of a survey of perceptions of threats and challenges in the Netherlands conducted for the

Organisation for Security and Cooperation in Europe (OCSE).[31] The conclusions were that:

1. there is no whole-of-government view of what constitute the medium-term and long-term threats and challenges;
2. ministries do not share whatever ideas they do have about them with other ministries; and
3. the whole-of-government view does not extend beyond its expected term of office.

The findings from this study correspond with those in a study of the central government's strategic capabilities by the Netherlands School of Public Administration (NSOB),[32] which also showed that strategy formulation is practised mainly within the confines of a ministry.

The structural integration of internal and external security – which still remain separate worlds despite the repeated observations about their interconnectedness – is no easy task and is largely *terra incognita*, but remains very important.

Assessment of the International Security Strategy

A Secure Netherlands in a Secure World was not the outcome of a regular process, but the result of a one-off exercise by officials from the Ministry of Foreign Affairs. It is unclear when this exercise might be repeated. The Strategic Monitor, the annual trend analysis by the Clingendael Institute and The Hague Centre for Strategic Studies (HCSS) (see also Sect. 4.3.3), as well as the expertise of the AIVD and MIVD, were used in drafting the International Security Strategy, which was sent to the House of Representatives after it had been approved by the Cabinet. It was unanimously adopted by the House of Representatives during a general meeting of the foreign affairs committee in October 2013.

A Secure Netherlands in a Secure World marked an important step forward. For the first time, the formulation of policy priorities had been preceded by an extensive security analysis. The International Security Strategy is also a policy document with a clear political narrative, containing typical Dutch ingredients such as multilateralism, promotion of the legal order, disarmament and arms control and the transatlantic alliance. But it also builds a bridge to the altered security environment, such as the consequences of climate change and the political and public concerns about them.

At the same time, scarcely any use is made of the possibilities of strategy formulation. The International Security Strategy refers to the need to set priorities, but does not specify what the Netherlands will no longer do, or will do less. The document sets out policy priorities, but does not formulate any specific objectives, results or deliverables. It also fails to mention the relationship with financial resources or how it will be embedded in a structured process that guarantees input from politicians, knowledge institutes and society. The strategy mentions new threats, but focuses on the traditional security agenda in setting policy priorities.[33]

Assessment of the National Security Strategy

When the National Security Strategy is assessed in terms of sound strategy formulation, a number of things stand out. A positive aspect is that it formulates a clear objective, i.e., the prevention of social disruption. Vital interests and core values are also identified. A formal process of risk assessment is also mentioned (see also Sect. 4.3.3). Another positive aspect is that it includes separate strategies specifically for cyber security and counterterrorism.

In light of the basic requirements of a strategic process (see Sect. 4.2), there are also some important shortcomings. There is no clear political role in the process of drafting the National Risk Assessments. The elaboration of scenarios, the risk assessment and the capacity analysis capacity are left mainly to experts in the National Network of Safety and Security Analysts (*Analistennetwerk Nationale Veiligheid*, ANV), with the government responding – at lengthy intervals – to the reports of their findings. The exclusive focus on internal security is a misjudgement in light of the connection between internal and external security. The National Security Strategy does refer to transnational threats such as terrorism, avian flu and the consequences of climate change, but the focus on internal security is clearly reflected in the scenarios that were produced, only a few of which address international developments. The same applies for the government's reactions to the reports of the findings, which, with just a few exceptions, devote scarcely any attention to the international situation.

4.3.2 The Whole-of-Government Approach

After the experiences with the 3D approach (Defence, Diplomacy & Development) in Uruzgan in Afghanistan, the Netherlands also accepted the 'integrated approach' in the International Security Strategy in 2013. It is one of the policy emphases for missions in fragile states or conflict zones:

> For an effective approach, it is important that the Netherlands establish the best mix of diplomatic, military and development instruments on a case-by-case basis. The government has various instruments at its disposal: diplomacy and political activities, use of the armed forces and the intelligence and security services, contributions to development cooperation activities, and efforts in other areas of governance, such as the judiciary and police.[34]

The Netherlands has also lobbied for the integrated approach within NATO and the EU and it now assumes an important place in NATO's Strategic Concept (2010) under the title 'comprehensive approach'. The approach is also a guiding principle for external action by the EU (see *The EU's comprehensive approach to external conflicts and crises*, 2013).

In the Netherlands itself, however, the whole-of-government approach to security has failed to keep pace with the vision of an integrated approach that is promoted internationally. Certainly, the changed security environment and the expansion of the security agenda have not been without consequences for the procedures of the Dutch government.

For example, the national crisis management structure has been radically altered, with stronger coordination by the National Coordinator for Security and

Counterterrorism (NCTV) and the National Crisis Centre (NCC) that has been established under his auspices. Formal consultation structures have been established at senior official and political level which can be mobilised in the event of a crisis (the Interdepartmental Crisis Management Committee and the Ministerial Crisis Management Committee, respectively).

Nor has the government stood still with regard to the integrated approach to international conflicts and crises. In the last decade cooperation between the relevant departments of the Ministry of Foreign Affairs, including those in the domain of the Ministry for Foreign Trade and Development Cooperation, and the Ministry of Defence has intensified. Other ministries, in particular the Ministry of Justice and Security, are now also involved. In contrast to the national security domain, there has been no extensive overhaul of structures or strengthening of the position of a particular ministry. The cooperation is highly operational in nature, with the focus on (possible) Dutch contributions to civil missions and military operations. There is more consultation between the ministries, with the senior civil servants in the Missions and Operations Steering Group coordinating both military operations and civil missions and liaising with the relevant cabinet members. There is no formal political body. The ministers of Foreign Affairs, Foreign Trade and Development Cooperation, Defence and Justice and Security are politically accountable for the specific input of their ministries and report jointly to the House of Representatives. The new budget for international security (BIV) also gives a boost to the integrated approach and interdepartmental coordination.[35]

The security structure in the Netherlands is characterised by a gap between forums (both political and official) that focus on international security, on the one hand, and bodies concerned with national security, on the other. This is an obstacle to adequate switching between policy and politics. In addition, the international 'compartment' appears particularly fragmented with numerous separate ministerial sub-committees and temporary commissions (for the decision-making on security aspects in the EU, the intelligence and security services, (special) operations and missions). This does not promote alignment.

Briefly, some important steps have certainly been taken to strengthen an integrated approach to the development and implementation of security policy, but they have taken place within the separate 'compartments' of internal and external security. There is no permanent, structured connection between the two at senior official or political level.

The strong interconnectedness of security, energy and the economy, of national and international security, and of Dutch policy and decision-making at alliance level (EU, NATO) was reflected at the time of the Ukraine crisis in 'a genuine balancing act between condemnation (sanctions), de-escalation and the safeguarding of the – mainly economic – interests".[36]

In the policy letter *Turbulent Times in Unstable Surroundings* (14 November 2014), the government acknowledged that the close interconnectedness of internal and external security called for effective coordination between the government agencies concerned with those issues. The cooperation is becoming 'increasingly intensive'. Nevertheless, in the existing constellation placing security issues in their context in relation to one another depends largely on individuals (ministers and civil servants) rather than on structures and procedures.

There are regular discussions in the Netherlands about whether to create a National Security Council – as a sub-committee of the Council of Ministers – to strategically manage an integrated security policy. The former member of the House of Representatives for the CDA, Jaap de Hoop Scheffer, called for the establishment of a National Security Council in September 2001 – shortly after 9/11. In 2004, a motion to the same effect was adopted by a majority in the House of Representatives, but it was not implemented.[37] Since March 2015, there has been a Ministerial Security Committee, chaired by the prime minister. In addition to the ministers of Foreign Affairs, Defence, the Interior and Kingdom Relations, Justice and Security and the deputy prime minister, the committee's members also include the National Coordinator for Security and Counterterrorism, the Director-General of the General Intelligence and Security Service, the Director of the Military Intelligence and Security Service, the Commissioner of the National Police and the Director-General for Political Affairs of the Ministry of Foreign Affairs. They meet every week to discuss matters of national and international security, make specific agreements on how to deal with issues that have arisen at national and/or international level and discuss whether a more in-depth study is needed of particular subjects or topics. The further research and decision-making on specific issues takes place in the Council for the Intelligence and Security Services (RIV).

Is the Netherlands 'too small' to have a National Security Council? Bailes describes the idea that small countries do not need formal structures or plans because the number of stakeholders is quite manageable and matters can be arranged decentrally, according to a careful (interdepartmental) balance of power or via individual politicians, as a typical weakness of the security policy of relatively small states.[38] In particular small developed countries which are, by definition, confronted with an overfull internal and external security agenda have to excel in strategy formulation, according to Bailes. There are no constitutional obstacles to establishing a National Security Council, but there is opposition to creating a new institution.[39] However, the complex and multi-dimensional security problems call for more horizontal coordination between ministers and departments.[40] The prime minister would ideally be responsible for ensuring the coherence of the policy within such a council.

4.3.3 Foresight and Risk Assessment

Foresight for the purposes of political decision-making in relation to international security is a relatively new phenomenon in the Netherlands. In 2010, the *Verkenningen: Houvast voor de krijgsmacht van de toekomst* [Future Policy Survey: A New Foundation for the Netherlands Armed Forces] was published on the initiative of the Ministry of Defence. This interdepartmental study presented four different future scenarios with four different answers to the question of how – and with what consequences – the world might develop in the coming decades. It then outlined four options for the use of the Dutch armed forces in the future. Each policy option emphasised a different function of the armed forces.

The authors of the study were convinced that the future scenarios were broadly applicable, not just within the Ministry of Defence but throughout government. They concluded that the scenarios presented in the *Future Policy Survey* could contribute – for example by establishing a common vocabulary – to the development of a whole-of-government vision and strategy. However, it did not come to that. Up to now the government has not again used what in 2010 was described as 'an inter-agency way of working that has broader possibilities for application within the government' from which 'valuable lessons learned ... can also be used in other policy areas'.[41] Nevertheless, the approach set out in the *Future Policy Survey* remains relevant, particularly the interdepartmental approach, the use of external experts, the drafting of scenario analyses and the development of policy options.

One positive aspect is the development of the Strategic Monitor. Since 2012, the Clingendael Institute and The Hague Centre for Strategic Studies (HCSS) have produced annual trend analyses for the ministries of Foreign Affairs, Defence and Justice and Security (in particular the National Coordinator for Security and Counterterrorism).[42] Meanwhile, at the request of these departments, the institutes also produce studies on specific subjects for the Strategic Monitor. The relevant departments are therefore able to make use of up-to-date analyses of the current situation in the policy development process.

Although the exercise of drafting an (externally oriented) *Future Policy Survey* has not been repeated, a National Risk Assessment (NRA) has been produced every year since 2007, see Fig. 4.1. The purpose of the NRA is to provide policymakers

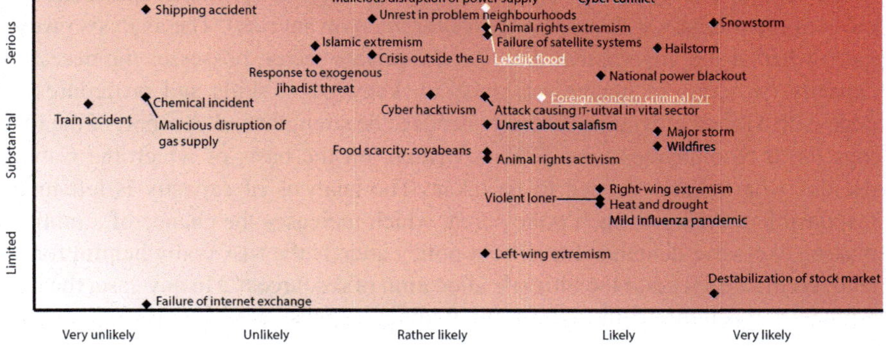

Fig. 4.1 The national risk diagram. (Source: National Risk Assessment, no. 6 (2014)

with insight into the relative likelihood and impact of different risk scenarios. This information is needed to define the capacities required and to set priorities to ensure that the Netherlands is optimally prepared for different types of disasters and threats. The NRA and accompanying scenarios are drawn up by the National Network of Safety and Security Analysts for the National Coordinator on Security and Counterterrorism.[43] The authors are independent, but the client has a significant voice in the choice of scenarios to be produced. The scenarios are integrated into the national risk diagram and arranged according to the likelihood of their occurring and their potential impact on society.

There are reservations to be expressed about the structure of the National Risk Assessment. In an advisory report on the so-called risk-rule reflex, the Council referred, among other things, to the complications attached to multi-dimensional risk comparisons and the technocratic nature of such an exercise.[44] There are, for example, reasonable doubts about the usefulness of the national risk diagram, which includes extremely diverse risks, ranging from black ice and snow storms to confrontations between individuals with a migration background and persons from the extreme right. How should this risk comparison be interpreted?[45]

A second reservation concerns the internal orientation of the scenarios. In themselves they provide valuable insights into possible events and their potential impact, but between 2007 and 2014 only three scenarios focusing on transnational risks were formulated. The value of the NRA for increasing understanding of the consequences of the interconnectedness of internal and external security is therefore limited.

A third reservation relates to the role of the National Network of Safety and Security Analysts. Identifying and weighing up national risks is an extremely complex process and is hedged by numerous uncertainties and gaps in our knowledge. Input is therefore essential from a variety of experts (including the National Institute for Public Health and the Environment (RIVM), the Research and Documentation Centre of the Ministry of Justice and Security (WODC), the AIVD and TNO). However, the NRA is not validated and verified by independent external bodies. Like anyone else, the experts concerned can be blinded by (unconscious) preoccupations and other form of bias.

Finally, the NRA is used to conduct a capacity analysis (primarily under the auspices of the ministry with responsibility for the relevant risk). The analysis investigates whether the government, but also the private sector, possesses the necessary capacity (in terms of manpower, materiel, knowledge, skills and procedures) to cope with a threat or whether capacity needs to be strengthened. A report of the findings (with recommendations) is then written, on the basis of which the Cabinet decides what measures need to be taken. The analysis of capacity is left to the responsible ministry in each policy area, which increases the chance of a ministry making excessive demands for its own policy area. Is the NRA really helpful for the analysis of capacity and the ultimate allocation of resources[46]? In any case, the government's reactions to the reports of findings suggest not. The *Voortgangsbrief Nationale Veiligheid* [Letter from the government to the House of Representatives with a progress report on the National Security Strategy] of 12 May 2015 does not even mention the National Risk Assessment 2014.

The National Risk Assessment is now published in amended form as the National Risk Profile (NRP),[47] in which five national security interests are distinguished: territorial security (the unimpeded functioning of the Netherlands as an independent state), physical safety (the unimpeded functioning of people in the Netherlands), economic security (the unimpeded functioning of the Netherlands as an effective and efficient economy), ecological security (the unimpeded continued existence of the natural living environment) and social and political stability (the unimpeded continued existence of a social climate in which individuals can function without being disturbed and groups of people enjoy living together within the benefits of the Dutch democratic system and values shared therein).

The National Risk Profile provides a comparative survey of risks (in terms of their likelihood and potential impact) ensuing from various disasters, crises and threats drawn up by National Network of Safety and Security Analysts. The first National Risk Profile in 2016 focused on potential disasters and threats that could disrupt *our* society, but devoted greater attention to transnational (geopolitical) threats and autonomous international developments. That is a step in the right direction.

However, in contrast to the annual National Risk Assessments, the National Risk Profile is only published every 4 years. In view of the rapid pace of developments, that represents a step backwards. It is also regrettable that, as a result, the States-General are not able to hold a political debate with the responsible ministers on the basis of an 'All Hazard' overview of risks every year.

The National Risk Profile also describes the capacity available to manage the risks. With regard to geopolitical threats, for example, this would be capacity for international cooperation (diplomatic, military, economic and development cooperation) and the capacity of the intelligence services and research institutes to provide information and analysis. The next step – a coherent analysis of capacity that identifies the capabilities that need to be strengthened and what is needed to strengthen them – also has to be taken, but falls beyond the scope of the National Risk Profile.

4.4 Extended Security Implies a Comprehensive Strategy

The greater interconnectedness and unpredictability of security issues have had a major impact on the thinking and actions of governments in relation to security in the last decade. Many countries have developed new strategic instruments, such as national security strategies and whole-of-government approaches, foresight studies and risk analyses. These instruments reflect an expanded, more comprehensive concept of security.

The Dutch government has also taken steps in the direction of 'comprehensive security'. On further reflection, however, it does not appear to have fully embraced the strategic instruments and the underlying philosophy. For example, the Netherlands has two separate strategic security documents, i.e., the National Security Strategy and the International Security Strategy. The whole-of-government approach to security issues, one of the articles of faith for international missions, is not applied consistently in this country. There is no overarching structure within which senior civil servants and ministers discuss the various aspects of security as a whole. Security issues are addressed in a fragmented fashion in various ministerial sub-committees and commissions and teams of officials that prepare policies. National and international security are compartmentalised.

Knowledge and anticipation are key words in dealing with complexity, dynamics and uncertainty, but the Dutch government possesses only modest and separate capacities for foresight and risk assessment. The attention to the preventive phase (including taking moderating measures) that precedes open armed conflict is not properly developed. The importance of mapping patterns in the 'geopolitics of emotion' is also not yet sufficiently recognised. The Clingendael Institute and the HCSS produce the Strategic Monitor every year and the National Network of Safety and Security Analysts produces the National Risk Profile, but there are reasonable doubts about the extent to which this knowledge actually benefits policy formulation. The long-term orientation and the match between strategic knowledge and policy are often inadequate. There is still a gap between science and policy in the area of foreign policy and defence.[48]

Last but not least, the Dutch security strategy and the strategy for the national armed forces derived from it are intrinsically connected with the security strategies of the alliances of which the Netherlands is a member. There is also a need for a more integrated approach in that respect, whereby the Netherlands must devote more attention to embedding it in the transatlantic alliance and the European Union's common security and defence policy. To put it bluntly, it is not only the Netherlands that determines its long-term choices in relation to its own defence efforts. International cooperation in the EU and NATO is inevitably at the expense of national sovereignty, but does increase these alliances' joint capacity to act.[49] Coordination with NATO's Strategic Concept (*Active Engagement, Modern Defence*, 2010) and the European Security Strategy is therefore absolutely essential. The EU's strategic framework dating from 2003 (*A Secure Europe in a Better World*) was in urgent need of revision. For example, the opening sentence read: "Europe has never been so prosperous, so secure nor so free". In June 2016, the High Representative of the Union for Foreign Affairs and Security Policy, Frederica Mogherini, presented a new global strategy entitled Shared Vision, Common Action: A Stronger Europe.[50] In short, a successful Dutch security and defence strategy is part of a *multi-level* strategy.

Endnotes

1 Cottey, A. (2013).

2 Franke, V.C. & Dorff, R.H. (2012); Jermalavicius T., Pernik, P. & Hurt, M. (2014).

3 Stolberg, A.G. (2012: 5).

4 Gray, C. (2014: 30).

5 Freedman, L. (2013: 609).

6 Freedman warns that without a convincing narrative capable of mobilising politicians and the public, there will be insufficient support for the use of military means in democracies or a serious risk of the erosion of what support there is. That risk is particularly great in countries like the Netherlands, with a political elite that is very sensitive to public opinion and a political system in which coalition governments are the norm. That automatically imposes demands on a strategic narrative (B. de Graaf, G. Dimitriu (eds.). *Strategic narratives, public opinion and war* (2015)). A strategic narrative has to meet endogenous and exogenous conditions. Endogenous conditions include a clear objective, legitimacy, the promise of success, consistency and a sound strategic policy. Exogenous conditions include a military culture, history and openness and tolerance (B. de Graaf, WRR brainstorming session, 28 January 2015).

7 Stolberg, A.G. (2012: 3).

8 Duyvesteyn, I.G.B.M. (2013: 4).

9 Freedman, L. (2013: 612).

10 Kramer, F.D. (2014).

11 Gray, C. (2014: 30).

12 Edmunds, T. (2012: 269–270).

13 Gray, C. (2014: 54).

14 Krepinevich, A.F. & Watts, B.D. (2009).

15 Bailes, A.J. (2014).

16 Bailes, A.J. (2014: 40).

17 The OECD explicitly makes a connection between a whole-of-government approach and complexity. A whole-of-government approach corrects systems that have moved too far into sector-based compartments and suffer from poor coordination and cooperation: “Such an approach, however, requires high-level guidance (e.g., the Centre of Government operating in a stewardship role) to set expectations and to ensure overall accountability, as well as cross-sectoral management, and cultural change” (OECD 2012: 4).

18 Jermalavicius T., Pernik, P. & Hurt, M. (2014).

19 WRR. (2014: 61–62).

20 Horton, A. (1999: 5–9).

21 Dreyer, I. & Stang, G. (2013).

22 Habegger, B. (2009).

23 Beck, U. (1986).

24 Power, M. (2004: 58–65).

25 WRR. (2011).

26 Cf. Williams, M.J. (2009).

27 Edmunds, T. (2012).

28 Corry, O. (2010).

29 The National Security Strategy distinguishes five forms of security: (1) territorial security. The unimpeded functioning of the Netherlands as an independent state in the wider sense, or the territorial integrity in a narrow sense; (2) economic security. The unimpeded functioning of the Netherlands as an effective and efficient economy. Economic security can be undermined if trade with an important international partner disappears, for example; (3) ecological security. The unimpeded continued existence of the natural living environment in and around the Netherlands; (4) physical safety. The unimpeded functioning of people in the Netherlands and its surroundings; (5) social and political stability. The unimpeded continued existence of a social climate in which individuals can function without being disturbed and groups of people enjoy living together within the benefits of the Dutch democratic system and values shared therein.

30 Colijn, K. (2016).

31 Haar, B. ter & Maas, E. (2014).

32 Wijnbelt, M. et al. (2013).

33 Haar, B. ter (2014).

34 International Security Strategy (2013: 17).

35 Matthijsen, C.J. (2014).

36 Hazelbag, L. (2015: 193).

37 Parliamentary Documents II 2003–2004, 27 925 no. 124.

38 Bailes, A.J. (2014: 40).

39 Hazelbag, L. (2015).

40 Bogdanor, V. (2005).

41 *Future Policy Survey* (2010: 11).

42 Recent versions are the Clingendael Monitor 2015, *Een wereld zonder orde?* [A World without Order?] and The Hague Centre for Strategic Studies, StratMon 2015, *The Return of Ghosts Hoped Past? Trends in Conflict and Cooperation*.

43 The National Security Strategy falls under the responsibility of the Minister of Justice and Security. Within his ministry, the National Coordinator for Security and Counterterrorism (NCTV) is responsible for the National Security Strategy and chairs the National Security Steering Group, which coordinates activities relating to national (internal) security. The Ministerial Security Committee meets every week to discuss national ('internal') security.

44 WRR. (2011: 41–43).

45 Nevertheless, risk comparison cannot be avoided from a policy perspective. See WRR (2014) with guidelines for a multi-issue risk and security policy.

46 Vlek, C. (2013).

47 National Network of Safety and Security Analysts (2016). See also Letter on Progress with the National Security Strategy, House of Representatives, 2014–2015, 30 821, no. 23.

48 Wiers, J. (2016).

49 For a discussion of the tension between undivided national sovereignty and shared sovereignty in international alliances, see Advisory Council on International Affairs, *Europese defensiesamenwerking. Soevereiniteit en handelingsvermogen* [European defence cooperation: sovereignty and the capacity to act], January 2012, pp. 10–13.

50 http://eeas.Europe.eu/top_stories/pdf/eugs_review_web.pdf; on this issue, see Antonio Missiroli (ed.), *Towards an EU Global Strategy. Background, process, references*, European Union Institute for Security Studies, 2015.

Chapter 5
Defence Policy in a Changed Security Environment

Ernst Hirsch Ballin, Huub Dijstelbloem, and Peter de Goede

5.1 Introduction

In the preceding Chapters (2, 3 and 4) we have discussed the changing nature of security and the expansion of the concept of security and explored their consequences for the formulation of strategy. In this and the following chapters, the consequences for defence policy and the armed forces are analysed.

The strategic trends that are decisive for the security of the Netherlands impose new demands on Dutch defence policy. To determine the type of defence policy required in a changing security environment, this chapter addresses the following questions:

- What are the historical background and points of departure of security policy and what defence policy has the Netherlands pursued in the last decade? (Sects. 5.1 and 5.2)
- What developments and trends are currently occurring in the international security environment? (Sects. 5.3 and 5.4)
- What consequences should they have for the main tasks of defence policy? (Sect. 5.6)

5.2 Background to Defence Policy

Beyond Entrenchment

For centuries, the Netherlands has had an international and European orientation that accords with the goals, interests and values of an increasingly urbanised and relatively prosperous trading country. The country has aspired, particularly since the second half

E. Hirsch Ballin (✉) · H. Dijstelbloem · P. de Goede
WRR, Scientific Council for Government Policy, The Hague, Zuid-Holland, The Netherlands
e-mail: ballin@tilburguniversity.edu; h.o.dijstelbloem@uva.nl; peter.goede@minbzk.nl

E. Hirsch Ballin et al. (eds.), *Security in an Interconnected World*,
Research for Policy, https://doi.org/10.1007/978-3-030-37606-2_5

of the nineteenth century, to an open, international economic system and a stable international legal order that curb arbitrary action and power politics by means of strong institutions, legal frameworks and the peaceful settlement of disputes. This is why the Netherlands was a driving force behind The Hague Peace Conferences (1899, 1907) and the establishment of the Permanent Court of Arbitration. It is aware that openness and interconnectedness are accompanied by dependence on the external environment and vulnerability to shocks originating elsewhere.

The idea that the Kingdom's security, and thus the security of our democratic society, is mainly a question of shutting ourselves off from the world around us has long been untenable. Although the general image of 'defence' is that of defending a fortress or a walled city, a new perspective was already emerging in the nineteenth century. Although colonial and trade interests played an important role in that respect, it is noteworthy that the Constitution of 1814 already instructed the Sovereign Ruler (the later King) "to maintain an adequate navy and army, recruited from volunteers, either indigenous or aliens, to serve inside or outside Europe according to the circumstances".[1] However, there was an important difference compared with the present situation, because protection of the national territory, and more specifically the province of Holland, was the military and mental fall-back position. From the end of the eighteenth century until 1940, a system of forts and areas of land that could be inundated if necessary formed the New Dutch Water Line (*Nieuwe Hollandse Warterlinie*), a line of defences designed to protect the cities of Holland and Utrecht against advancing hostile armies.[2] This view of defence as entrenchment against external forces – in the same way as dykes protect against encroachment by the sea – was a response to the characteristic feeling of most people in this country, which was described by the sociologist Weidenhaus as 'concentric-linear': life unfolds in the same region in a succession of similar experiences. The nineteenth-century nationalism and the preference for forming culturally and economically homogeneous nation states, separated by borders from other states, reflected that view.[3]

Towards the end of the nineteenth century, however, a change started to occur in the thinking about international relations. This transformation of the world (as the title of Osterhammel's book in 2011 puts it) set in motion a trend where a growing number of people, some driven to it by their circumstances, became less bound to a single place but lived their lives more in networks and as a succession of episodes. This led to tremendous growth of the transport sector, migration and international trade.[4] The Netherlands' trade interests also prompted a change in perceptions of international relations, which was partly, but certainly not exclusively, connected with colonial interests; the international regulation of shipping on the Rhine was also an important subject of Dutch diplomatic activity. The understanding that trade and transport were best served by peaceful international relations was an important driving force of Dutch foreign policy, which was manifested, among other things, in the preparation and organisation of The Hague Peace Conferences in 1899 and 1907, which resulted in treaties on the peaceful resolution of international disputes and on the restrictions to be observed in war. With these treaties, the first steps were taken in moving away from defending oneself by halting hostile armies towards preventing the need to do so.

Active Membership

The still fragile development of treaty-based international relations suffered a dramatic setback with the First World War. New efforts to revive them, such as the General Treaty for Renunciation of War as an Instrument of National Policy (the Briand-Kellogg Pact) signed in Paris in 1928, also collapsed when Nazi Germany and the imperialistic Japan again launched wars of aggression. In the first decade after the Second World War there were fears of a Third World War, potentially even more disastrous because of the availability of nuclear weapons.[5]

The Kingdom of the Netherlands, which until the Second World War had tried to remain safe through a policy of neutrality, immediately joined NATO after the war, while in its foreign policy it played a prominent role through diplomacy and aid for the Third World. The amendment of the Constitution in 1953 reflected this, for example, with the instruction to the government to promote the development of the international legal order (currently Article 90). The Dutch armed forces (which were made up in part of conscripts) made a qualitatively and quantitatively significant contribution to the alliance. NATO and the Warsaw Pact, with disproportionate contributions from the two superpowers, armed themselves to be prepared for the worst and developed military strategies for a war in which the North German plains would be an important battleground.[6]

Since the 1950s, the Netherlands has protected itself through its active membership of the UN institutions, NATO, the EU and other regional European institutions such as the Organisation for Economic Cooperation and Development (OECD) and the Council of Europe, and later the Organisation for Security and Cooperation in Europe (OSCE). The UN institutions provided the platform for peaceful resolution of international disputes, multilateral cooperation and protection of human rights.[7]

The threatening crises over Berlin and Cuba at the beginning of the 1960s were de-escalated just in time.[8] In the ensuing years, the threat of war gave way to an international security environment in which stability was derived from mutual deterrence and – on balance – adequate means of communication, symbolised by the UN Security Council and the 'telephone hotline' established between Washington and Moscow, to prevent wars from starting by mistake. In addition to the availability of forums for diplomatic negotiations and the resolution of disputes, the realisation in the United States (with its NATO allies) and in the USSR (with the Warsaw Pact countries) that nuclear annihilation would not produce any winners in a world in which countries all depended on one another ultimately also had a salutary effect.

The policy of détente (from the 1970s) and agreements on arms control, followed by the implosion of the communist power structures, heralded the start of a new era at the end of 1989. With the United States as the sole remaining superpower, the expectation quickly spread that that country would be the guardian of international peace. Euphoria was the dominant emotion and that gave a powerful boost to the EU. Although the savage civil wars in the former Yugoslavia and military conflicts in the Caucasus disturbed the picture of a peacefully reunited Europe, optimism prevailed. Defence spending was substantially reduced throughout Europe; in the Netherlands by 25% between 1990 and 2014, a figure that is not corrected for the growth of GDP.

As a percentage of GDP, expenditure fell from 2.5% to 1.2% (see Chap. 6). National service was suspended in the Netherlands in 1997. The total number of personnel in the armed forces declined by three-quarters, not even counting the disappearance of the reservists.

In the context of this Pax Americana,[9] the Dutch armed forces no longer concentrated on defending the territory of the Western European allies against attacks from the East – it was assumed that this threat had disappeared for good – but on participating in military actions in other countries, usually under the title of ‘crisis management and peacekeeping operations’, under the auspices of NATO, the UN or the EU. This task has gradually evolved in the sense that human rights and the democratic rule of law are not seen as the preserve of this country’ society, but are also an entitlement of peoples elsewhere in the world.[10] The collective defence under American leadership in NATO protected the Netherlands against the threat from the Soviet Union and spared international relations from ‘continental’, Franco-German dominance. And the European Community gave the Netherlands access to a stable internal market with the other member states, an economic level playing field and the growth of prosperity which, with the establishment of its own welfare state, acted as a buffer against global economic shocks.

In 2000, the article of the Constitution devoted to the tasks of the armed forces, which were previously exclusively the defence and protection of the interests of the Kingdom, was supplemented with ‘to maintain and promote the international legal order’ (Article 97), together with the obligation for the government to inform the States-General in advance if the armed forces are to be deployed or made available to maintain or promote the international legal order (Article 100).

The Dutch armed forces have participated in approximately 50 international military missions since 1990. The structure of the streamlined Dutch armed forces was tailored to these types of operation, with a fairly wide range of military capabilities, but with scant possibilities to carry out the operations for more than a few years. During this period the image of the ‘Swiss army knife’ was introduced to symbolise the versatility and flexible deployability of the armed forces.[11] The symbol of the Swiss army knife was intended to express the fact that a range of instruments had been combined in a single toolbox, but that each could be taken out and used as required. In that context it was assumed that the Netherlands “(is) an open and prosperous country in a safe region” (p. 298) and that the changes since the 1990s could be extrapolated into the future (p. 303). As provided for in the Coalition Agreement in 2010, the Minister of Defence at the time wanted to remain “as close as possible” to this concept, but with the reservation that this was “very risky” as a result of the continuing cutbacks.[12]

More than a quarter of a century after 1989, the observation has to be made that the expectations of peace and stability have been rudely shattered. The current situation of shortages and vulnerability in the Dutch defence forces has to be attributed to the fact that even long after the signs that times had changed, in other words even after 2008, policy continued to build on the expectation that serious military conflicts would pass us by. H.A. Winkler[13] describes how the triumph of the Western model of society ended in tragedy with the terrorist attacks of 11 September 2001. The succeeding period was devoted to the war against terror, followed by the global financial crisis in 2008. In the ensuing years Obama's America was evidently overburdened, according to Winkler,[14] while the problems accumulated for the EU and the West: the debt problem, the demise of the Arab Spring, the growing ambitions of Russia and China, the Ukraine crisis and, since the crisis year of 2014, a 'globalisation of terror' (p. 549).

The coinciding of the financial crisis with the radical change in the international security environment led, even after 2008 and right up until 2015, to a continued reduction of spending on defence by the Dutch government, but also of spending for conflict prevention through (adequate forms of) development cooperation. This determines the complexity and urgency of finding a new frame of reference for Dutch policy. The finding from the Ministry of Defence's foresight studies that security policy must be equipped to meet multiple challenges was in itself correct. The dramatic deterioration in the outlook after 2001, and even more so after 2014, confirms that the old model of security behind a territorial line of defence is no longer fit for purpose. Insecurity presents itself in the same continental and intercontinental networks that characterise our lives and societies.

5.3 International Operations and Multilateral Frameworks

The points of departure and developments outlined in the previous section, the reactions to them and the lessons that have been drawn from them are still apparent. The Netherlands makes an active contribution to international crisis and stabilisation operations (Sect. 5.3.1) and invests in more intensive cooperation within the tried and trusted multilateral frameworks (Sect. 5.3.2), but today it does so with fewer resources (financial and otherwise) than in the 1980s and 1990s.

5.3.1 Contributions to Crisis and Stabilisation Operations

When the Cold War ended, the classical threat of aggression against the country's own territory was no longer regarded as germane. Since then the Netherlands' ambition has been to make an active contribution to crisis and stabilisation operations with the aim of promoting the development of the international legal order, addressing the fundamental causes of conflicts and translating its own military performance into political influence. Since 1990, the armed forces have been transformed into a flexible expeditionary force in a high state of readiness.[15] Conscription was suspended and the Netherlands' professional soldiers started participating in numerous peacekeeping and other missions.[16,17]

Since then the relevance of the armed forces is no longer determined by the extent to which they are not deployed (deterrent), but by the frequency with which they are sent out on missions.[18] The question is whether that should be the case, since it has an impact on defence planning. The central question is no longer what the Netherlands needs in terms of defence capabilities in light of the security situation, but what political 'ambitions' – expressed in numbers, scale and intensity of troop deployments – are formulated in the Netherlands and how the armed forces can achieve those ambitions with diminishing financial resources.[19]

The participation in these operations also resulted in agreements on the political decision-making in the form of the Assessment Framework (1995) and the new Article 100 that was inserted in the Constitution in 2000. That article provides that "[the government] shall inform the States-General in advance if the armed forces are to be deployed or made available to maintain or promote the international legal order. This shall include the provision of humanitarian aid in the event of armed conflict."

The political support for the invasion of Iraq and the contribution to NATO's ISAF mission in Afghanistan made it clear, however, that in practice the lengthy list of criteria for assessing the desirability and possibility of participating in missions in the Assessment Framework provided no guarantee of the legal legitimacy of Dutch military action or of political support for it. Outspoken objections and counter-arguments were not sufficiently considered and too often the expertise of specialist academics was left neglected. In its report on the intervention in Iraq, the Davids Committee referred explicitly to the importance of an international-law mandate.

The Assessment Framework has therefore also been amended a number of times. The most recent change came in 2014, when the importance of protecting the civilian population and providing care for soldiers after a mission were added as issues that needed to be considered.[20] Nevertheless, there is still an area of tension between the ambition of retaining international influence and relevance by contributing to international operations, the objective of strengthening the international legal order and the aim of retaining political and social support for that ambition. This has been illustrated once again by the debate about the bombing of Da'esh in Iraq and Syria. Only if such actions are embedded in a widely-supported international long-term strategy for reconstruction, which also includes protection of the civilian population and the prevention of new conflicts, might this tension possibly diminish.

5.3.2 Further Integration with Multilateral Frameworks and Bilateral Partners

There are various underlying reasons for the further integration of the Netherlands' efforts in the field of security with those of its allies: the Netherlands' declining political weight since the enlargement of the EU and NATO, the growing significance of the EU in the areas of justice and home affairs and foreign and security policy, the waning commitment of emerging powers to multilateralism, the declining American willingness to pay the costs of European security and the Netherlands' wish to cut spending on foreign and security policy. 'Peace without money, war without Americans' – that is the dual challenge facing countries in Europe, especially since Donald Trump became President of the United States.[21]

That challenge has to be faced in an altered landscape of conflicts. As Fig. 5.1 shows, there are few conflicts *between* states and internal problems within countries also do *not* seem to have increased in the last 20 years. The intensity of conflicts seems to be diminishing, see Fig. 5.2. The latent threats have changed, however, and that calls for a revised defence policy with a different role for the armed forces.

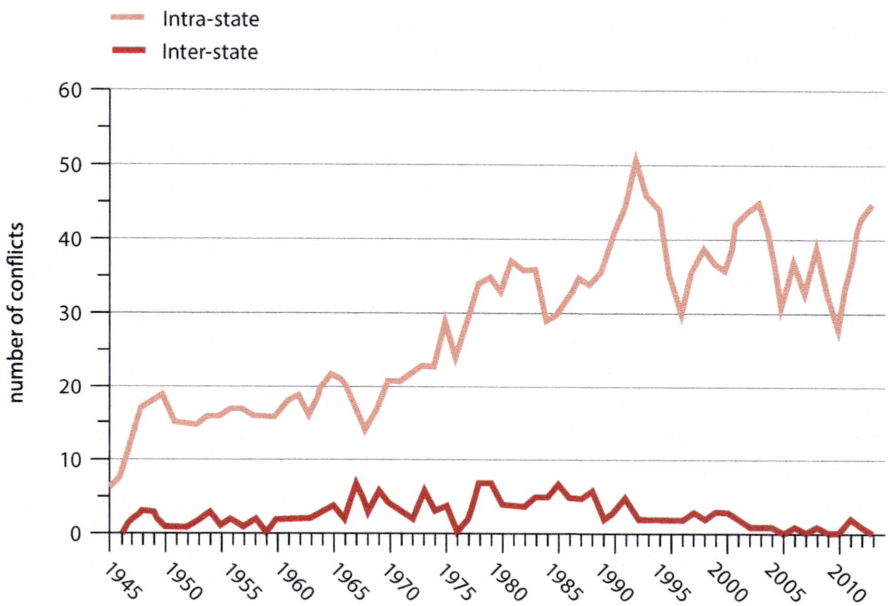

Fig. 5.1 Intra- and inter-state conflicts 1945–2013. (Source: Clingendael Strategic Monitor 2014)

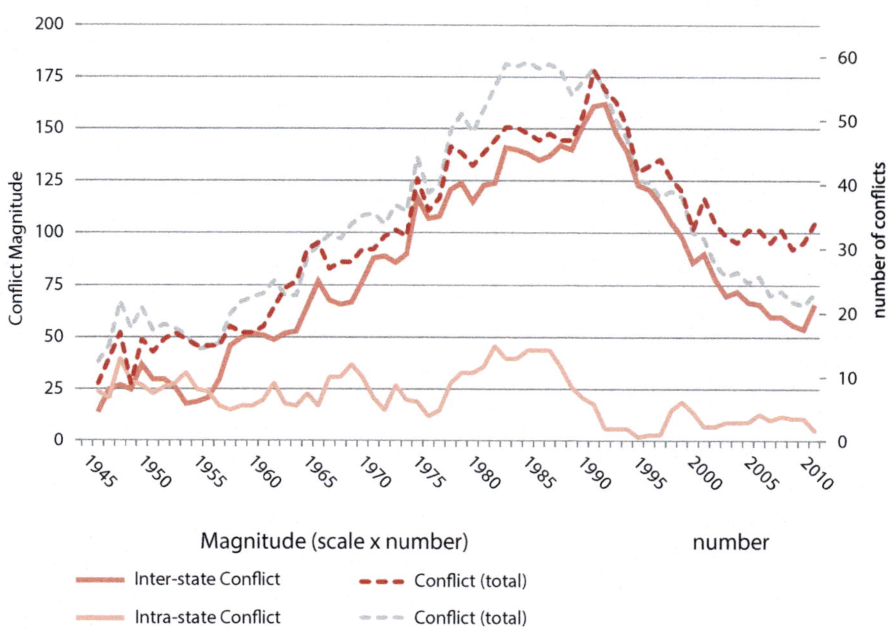

Fig. 5.2 Trends in conflicts: 1946–2012, total magnitude (number and scale) and numbers of inter-state and intra-state conflicts worldwide. (Source: HCSS Strategic Monitor 2013: 15)

For some time, Dutch policy has been a 'policy of contribution'; the Netherlands never undertakes international missions alone, but is asked to participate, and can use that position to exert political influence.[22] Consequently, there is also a growing realisation that the further consolidation and sharing of essential capabilities is imperative, but also creates mutual dependencies and vulnerabilities. After all, the decision-making in NATO and the transition from a European Security and Defence Policy (ESDP) to a genuine Common Security and Defence Policy (CSDP) are intergovernmental processes, depend on differing and often conflicting – also in domestic political terms – agendas of the member states and proceed slowly and with difficulty. In that constellation, countries like the Netherlands are relatively vulnerable to the attitude of the larger countries.[23]

The Netherlands also supports the growing practice of civil-military coordination in UN and NATO operations; the international centre of expertise on this subject is based in the Netherlands. In this practice, military units coordinate their actions and the necessary support with the civil parties on the ground.[24] The UN already appealed in 1992 for intensive efforts to ensure that crisis management operations made a genuine contribution to enduring peace and security. There have been numerous publications since then containing calls for more coherent approaches to security and development. In 2003, the Netherlands developed its own policy framework based on the principle "as civil as possible, as military as necessary".[25] The aim of civil-military coordination is to support the peace process and security, win the trust of the local population and – where necessary – repair the infrastructure and temporarily perform administrative and policing duties on a modest scale. An interdepartmental consultative structure was created for these subjects, in which the ministries of Foreign Affairs (including the Directorate-General for International Cooperation) and Defence participate, and other ministries and NGOs attend by invitation.[26]

In addition to the track focusing on the EU, NATO, the UN and the OSCE, the Netherlands also follows a bilateral track with countries including Germany, Belgium, France and the United Kingdom. It also explores the possibilities for closer cooperation in coalitions of the able and willing. The Netherlands pushes for closer bilateral cooperation with Belgium, France, the United Kingdom and Germany. Examples of the steps it has taken are the gradual deepening of the army's integration with the German *Bundeswehr*, the intensive cooperation with Belgium in naval affairs and protection of airspace and the closer operational cooperation with France arising from the Netherlands' contribution to the EU mission in Mali (MINUSMA).

The European Union's Common Foreign and Security Policy (CFSP) has become a fixture of Dutch security policy.[27] It reflects various cornerstones of Dutch policy, particularly the need for an 'integrated approach' in order to enhance the cooperation between the various actors in the EU and NATO, the desire to strengthen Europe's military capabilities, for example through pooling and sharing as is done at NATO level. Another urgent issue mentioned is the need to improve political decision-making within the EU on rapid reaction units (the EU Battlegroups) and to involve national parliaments in those decisions.[28] The Netherlands also contributes to various EU civil and military missions.

At present Dutch security policy does not assign the EU a significant role beyond the domain of the Common Foreign and Security Policy, with the exception of the maritime security strategy, the European defence industry and, in that context, the provision of funds for research into the use of dual-use capacities for dual use, in other words for civil and military use. Some areas of common ground have arisen, however, such as those between migration and border control and between development cooperation and security, in the African Peace Facility for example. At the European Council meeting in June 2015, the Netherlands again stressed the importance of better and more systematic European defence cooperation and of strengthening partnerships with the UN, NATO, the OSCE and the African Union.[29]

The experiences in Libya, Mali and the Central African Republic and the crisis in Ukraine demonstrate that the member states lack the political will to use the EU for crisis management. Progress in jointly tackling the shortfalls in European military capabilities via pooling and sharing of national investments and expanding the European security agenda to policy areas outside the Common Security and Defence Policy is also advancing very slowly. In December 2013, the European Council called on the High Representative to present new proposals for effective cooperation. Since then, the European Maritime Security Strategy (June 2014), an EU Cyber Defence Policy Framework and a Policy Framework for Systematic and Long-Term Defence Cooperation have been adopted, but these initiatives are still in their infancy.[30] The strategic challenges in the regions surrounding the EU will certainly also require changes in the European Neighbourhood Policy (ENP), which has proved ineffective and should provide for differentiated relationships with the countries in the Middle East and North Africa and a more extensive toolbox of instruments, including capacity for a migration strategy and a rapid reaction force based on the Common Security and Defence Policy. At the European Council meeting in June 2015 it was decided that the EU will continue to develop "an effective, visible and result-oriented Common Security and Defence Policy, further develop both civil and military capabilities, and strengthen Europe's defence industry".[31]

The EU's High Representative subsequently prepared a strategic vision of the EU's role in the world. After an extensive consultation process, the Global Strategy for the EU's Foreign and Security Policy was published in June 2016 (*Shared Vision, Common Action: A Stronger Europe*). The strategy calls for strengthening of defence cooperation in Europe. The presentation of the Global Strategy coincided with the referendum on the United Kingdom's EU membership. With Brexit, the United Kingdom will also find itself 'outside' the Common Security and Defence Policy, which will complicate efforts to intensify European defence cooperation. Britain's departure from the EU would seem to increase the chances of a more European, but less military Common Security and Defence Policy.[32] Nevertheless, there is still a lot of uncertainty about how Dutch policy should react in this context without the United Kingdom as a permanent partner.

The priorities in the Global Strategy will be translated to the Common Security and Defence Policy.[33] The *EU Implementation Plan on Security and Defence* contains proposals for deepening defence cooperation, for example through the use of so-called 'permanent structured cooperation' (on the basis of Article 42(6) of the Treaty of Lisbon; for more on – reform of – the PESCO mechanism, see also CEPS 2015). A proposal for a European Defence Fund was unveiled on 30 November 2016. On 11 December 2017, 25 Member States committed to the activation of PESCO by means of a Council Decision.

Since the difficult missions in Afghanistan, and particularly the growth of instability on the eastern and southern flanks of the EU, NATO has been engaged in a reorientation to 'essential core tasks', including the collective defence of allied territory, as well as global military crisis management and security cooperation.[34] The process is not proceeding without problems. After all, the Ukraine crisis also exposed the divisions within NATO: over the relative importance of the threats on the eastern and southern borders in the long term, over the most urgent investments in the military forces (flexible, mobile troops or rather a permanent force stationed on the border), over the feasibility and desirability of closer military cooperation and integration of troops.[35] This will also come to the fore in the forthcoming review of the Strategic Concept from 2010 ('Active Engagement, Modern Defence') and the new *Comprehensive Political Guidance*, which sets out the priorities for the next cycle of NATO's planning process.[36]

The NATO summit in Wales in September 2014 underlined the renewed importance of collective defence (c.q. Article 5 of the NATO treaty, which provides that an attack against one member state shall be considered an attack against all of them and that all member states will cooperate in repelling the attack). For example, the Readiness Action Plan (RAP) to improve the deployability of allied troops was adopted. The plan encompasses a reorganisation of the NATO Response Force (NRF), including the establishment of the Very High Readiness Joint Task Force (VJTF), comprising a so-called rapid reaction force of 5000 troops.

The allies also committed themselves to raising defence spending towards 2% of GDP, decided to intensify NATO patrols of the airspace over the Baltic states and announced large-scale military exercises along the eastern border. The Netherlands, together with Germany and Norway, is supplying ground troops for the VJTF. The Netherlands also signed a declaration of intent in 2013, together with the United Kingdom, Denmark, Estonia, Latvia, Lithuania and Norway, concerning participation in the British initiative for a Joint Expeditionary Force (JEF) capable of responding quickly to crises without having to rely on decision-making in all 28 member states. Allies can decide on a case-by-case basis whether to contribute to the modules to be deployed. The Netherlands chose to contribute to maritime security with a joint British-Dutch amphibian force.

At the NATO summit in Warsaw in July 2016, decisions were made on the forward deployment of units in the Baltic states, Poland, Romania and Bulgaria. This 'trend' could continue in the coming years (and have consequences for the Dutch armed forces), depending on developments in relations between the US/NATO and Russia. Also indicative of new threats to the collective security of NATO is the Cyber Defence Pledge, which was adopted at the NATO summit in Warsaw, in which cyberspace was recognised as a domain in which NATO has to defend itself in the same way as the allies defend themselves on land, at sea and in the air. In principle, therefore, a cyber attack can be equated with an attack with conventional weapons (and therefore activate Article 5).

5.4 The Security of the Caribbean Parts of the Kingdom of the Netherlands

When one speaks of Dutch defence policy, thoughts normally turn only to that part of the Kingdom of the Netherlands that lies in Europe. But that is an incorrect constraint. The Charter for the Kingdom of the Netherlands leaves no doubt about that. International relations and defence encompass the entire Kingdom, not just the Netherlands or the European part of the Netherlands. The policy on international security extends to the entire state in the international law sense, in other words also those parts of the state situated in the Caribbean region.

However, international relations and defence as they relate to the Caribbean parts of the Kingdom do possess specific features. In the first place, there are differences arising from geography. The distance from Europe is too great and the scale of the relevant islands is too small to organise military activities there in the same way as in the European part of the Netherlands. Articles 30–33 and Article 35 of the Charter regulate the contributions of the countries of the Kingdom in terms of the staffing, materiel and financial requirements of defence. Article 34 grants the government of the Kingdom the power, in the event of war or the threat of war or a threat to or the disturbance of internal order and peace, to declare any part of the Kingdom to be in a state of war or a state of emergency with a view to maintaining internal or external security. This provision cannot be applied, however. The national law that should regulate it[37] has never been passed and a proposal in 1994 to rectify the situation has still to be implemented.

It is also relevant that Article 6 of the North Atlantic Treaty limits the obligation of collective self-defence to the territory north of the Tropic of Cancer. Although the obligation for the EU member states to provide mutual assistance in the event of armed aggression (Article 42(7) of the Treaty on European Union (TEU)) does extend to the ultra-peripheral regions of the EU south of the Tropic of Cancer, such as the French Antilles, it does not apply for the so-called overseas countries and territories, in other words those parts of the member states to which, in principle, EU law does not apply by virtue of Article 52 TEU and Article 355 of the Treaty on the Functioning of the European Union (TFEU).[38] The Caribbean parts of the Kingdom of the Netherlands currently fall into the latter category, although there is a possibility – in particular for the islands falling within the Dutch state system, Bonaire, Saint Eustatius and Saba – that this will change in the near future. Although the Caribbean parts of the Kingdom lack the protection of NATO and the EU under international law, their defence is *de facto* embedded in alliances with the US and the two other European states with Caribbean territories, France and the United Kingdom. These countries coordinate the coastguard's activities to combat drug trafficking and by special agreement the US has an air force support base on Curaçao.

In addition to these differences arising from geography and from treaties, there are also differences in the nature of the security issues. The Caribbean is a part of the world with a large number of island states and regions that differ from one another in various respects. The changes in international relations in the Caribbean region in the last century were caused by decolonisation and revolutions, not by wars aimed at territorial expansion at the expense of another state. Naturally, this does not mean that the region is assured of lasting peace. Since the Argentine-British war over the Falkland Islands, the question is raised from time to time of whether Venezuela will always respect the sovereignty of primarily European governance of three nearby islands.[39] It is not inconceivable – as was the case in Argentina at the time – that domestic unrest will spill over into a foreign adventure. For its part, Venezuela might, in the event of serious disturbances on one of the islands, feel called upon to protect its interests in the oil refining on Curaçao and Aruba by military means.

In other words, as in Europe and around the Mediterranean, internal and external peace are interconnected in this part of the world. Placing the connection between external and internal peace on the agenda is complicated, however, by the fact that, with the exception of the 'guarantee function' and ensuring compliance with treaties, the Government of the Kingdom has no general powers in internal matters. Nevertheless, the binding nature of treaties on the Kingdom is important when it comes to the essential features of a free and socially just society. When internal peace is disturbed by clear violations of civil, political, economic, social or cultural rights – the subjects covered by, among other things, the principal UN conventions – the organs of the Kingdom are obliged to defend the citizens of every part of the Kingdom. The same applies for other possible causes of social disruption, such as major corruption, drug trafficking and other forms of organised crime. Those issues also involve treaty obligations of the entire Kingdom – but the responsibility of each individual country must always take priority, both for constitutional reasons and in the interests of the political cohesion within each of the societies. An excessive inclination to intervene could itself become a source of tension.

Little research has been conducted into the special features and requirements of security policy in the Caribbean region. Twenty years ago Ivelaw Griffith investigated the unique security problems in the Caribbean in a study for the American Institute for National Security Studies (part of the National Defence University). The region is vulnerable partly because of its fragmentation in a constitutional sense: 'traditional concepts of sovereignty cannot cope with torrential trans-border flows of narcotics, money, arms, and immigrants'.[40] In addition to a number of unresolved border disputes, the most important of which involves Venezuela and Guyana, major risks arise from the 'geonarcitics', the term Griffith uses to denote a multidimensional phenomenon with "drug production, consumption-abuse, trafficking, and money-laundering" as the principal problems and joint actions by the Caribbean entities in response to them.[41] He refers to the need to strengthen the stability of democratic governance, in conjunction with protection of human rights and a properly functioning judicial system (pp. 69–72). Mutual cooperation is needed for survival, but is not enough; help from outside – he mentions the US in that context, but France, the United Kingdom and the Netherlands could also feel they are being addressed – is also essential (pp. 73–74). Twenty years later, these observations have lost none of their actuality, but would now have to be supplemented with the subject of migration, including the criminal abuse of migrants in the form of human trafficking. Like drug trafficking, human trafficking is accompanied by other forms of crime (money laundering, corruption and intimidation of public officials).

The wide-ranging studies by Prevost, Vanden et al. in 2014 are also concerned with Central and South America. In relation to the drug trade, they refer in particular to a loss of sovereignty in regions where armed gangs (*maras*) and large-scale criminal organisations have assumed power.[42] They too advocate external support for fragile states in order to contain the risks arising from failing states and uncontrolled regions (pp. 187–188). Furthermore, a situation could arise in this part of the world where large numbers of people migrate because of civil wars, pandemics or climate change and so further undermine the fragile stability of the Caribbean political entities.

When one compares the experiences in the last few years with the findings from this research, what stands out is that they point in the same direction. In the fragmented Caribbean environment, security policy has to be implemented on the basis of cooperation aimed at creating stability. The Dutch military component can only function in a supporting and supplementary role alongside those of the Caribbean states and other entities, the US, France and the United Kingdom. But that involvement of the Netherlands is essential, given the relatively great importance of the islands connected to the Netherlands in this part of the world. Strengthening the stabilising institutions of the Caribbean countries and island territories in the Kingdom, such as the judiciary, the public prosecution service and politics, must remain a high priority. Stronger socio-economic development ultimately offers the best counterweight to the creation of power vacuums; these can also be no-go areas in slums or sectors of the economy without effective supervision.

The Kingdom's armed forces will have to continue playing a role in maritime security. Although there are no complaints of shortcomings on the part of the Royal Netherlands navy at the moment, its capacity is also under pressure in the Caribbean region. Capabilities to protect against threats to flow security will also demand attention – given the risks associated with abuse of vital transport routes –from the perspective of security in the Caribbean parts of the Kingdom.

5.5 Developments and Trends in the International Security Environment

Defence policy is emphatically confronted with the deep-seated changes in the security environment. As a result of the convergence of 'old' and 'new' threats to security, together with the greater interconnectedness of internal and external security, the negative consequences of globalisation are appearing alongside traditional geopolitical threats. Complex global issues, such as climate change, migration and the crumbling liberal international order based on laws and treaties, demand attention. The same applies for the growing rivalry between (groups of) states over issues such as spheres of influence, open spaces (the sea, airspace and space), scarce natural resources (including water, energy and raw materials) and markets. Non-state enemies have profited from globalisation and become increasingly important actors. Security is no longer primarily territorial or static, but shifts within and between networks, both physical and digital. A great deal of research is therefore being conducted into the impact of these developments on the future deployment of armed forces. NATO, for example, is drafting future scenarios for developments such as the use of and denial of access to regions, large-scale disasters due to climate change, weapons of mass destruction and cyber threats (interview with ACT NATO, February 2015).

What developments and strategic trends are likely to determine the security environment in the Netherlands and Europe in the coming decades? The analyses performed by various think tanks (including the Centre for European Policy Studies, the Clingendael Institute, The Hague Centre for Strategic Studies, the *Institut des Relations Internationales et Stratégiques*, the European Institute for Security Studies, the *Stiftung Wissenschaft und Politik* and the German Marshall Fund) can provide an initial impression.

The Eastern Flank of the EU

The states on the EU's eastern flank are high on the list of potential flashpoints of future instability and conflict, because of their weak institutions, imbalanced economic development, corrupt political elites, ethnically diverse populations and arduous leadership changes. These states are expected to become more exposed to the disrupting influence of Russia. Popular revolutions such as the Rose Revolution in Georgia in 2003, the Orange Revolution in Ukraine in 2004, the failed Tulip Revolution in Kyrgystan in 2005 and the Arab Spring have awakened Russian fears of growing political 'contamination' from the Western world. Since the war with Georgia (2008), the orientation of the Russian Federation has steadily shifted to its own multilateral institutions, such as the Eurasian Union and the Collective Security Treaty Organisation. In this way, the country is claiming strategic leadership of the Central Asian region.[43] Russia's annexation of Crimea in March 2014 illustrates the fact that hard military confrontations have not disappeared, even in the current era of globalisation, greater mutual dependencies and indirect exertion of power.[44] There is considerable debate among experts about whether the Russian action in Georgia and Crimea – involving, among other things, a combination of diplomatic pressure, propaganda and interventions by special troops and local activists – marks the start of an era of less inter-state trust with hybrid threats and warfare.[45] Russia's military and economic strength in the longer term is also debated. But there is little doubt any longer that 'Ukraine' is sending a warning.

The Southern and Eastern Flanks of the EU

The instability and potential for conflict along the southern and eastern flanks of the EU will persist in the coming years. With the descent of the popular uprisings in the Arab Spring in 2011 into large-scale destabilisation and refugee crises, the Middle East and North Africa (MENA) region is experiencing a turbulent period. Europe is feeling the effects of that in the form of an influx of refugees and irregular migrants, which has caused a severe escalation of social and political tensions within and between EU member states. The instability on the southern border seems set to continue for the time being. With the exception of Oman and Morocco, the human rights situation has deteriorated in all of the MENA countries and their governments are under pressure from economic shocks and enkindled terrorism. The south of Algeria, Libya, the Sinai region of Egypt, parts of northern Iraq and Syria and Yemen all face acute security problems because terrorist and criminal networks have stepped in to fill the vacuum left by the complete or partial absence of central authority. The civil war in Libya is feeding terrorism, drug trafficking and people smuggling and destabilising the global oil market. The death toll in the civil war in Syria has climbed into the hundreds of thousands and the war will probably set the country's economy back by several decades. The stream of Syrian refugees into neighbouring Lebanon also increases the risk of internal uprisings in that country.

The Sahel countries and the Horn of Africa (Senegal, Gambia, Mauritania, Mali, Burkina Faso, Niger, Nigeria, Chad, Sudan, Ethiopia, Eritrea, Somalia and Djibouti) form a region of growing instability (see Figs. 5.3 and 5.4). The relative vulnerability of each country is indicated by the depth of the colour in Fig. 5.4 which is based on the State Fragility Index (SFI). These countries are characterised by rapid population growth and suffer from recurring cycles of violence and weak governance.

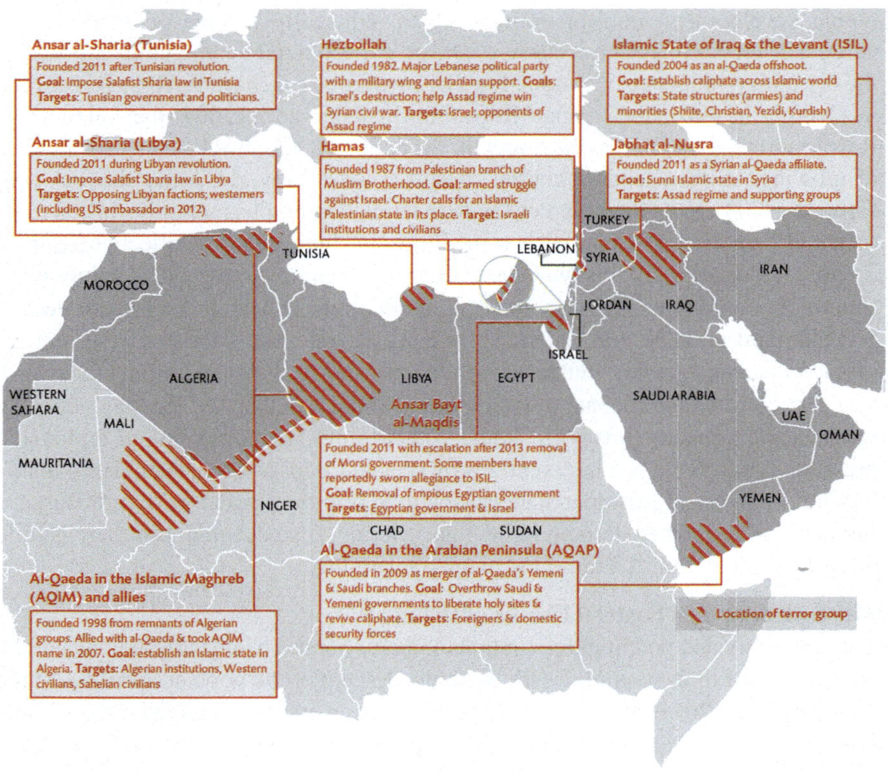

Fig. 5.3 Terrorist groups in the Middle East and North Africa. (Source: Missiroli et al. 2014: 58)

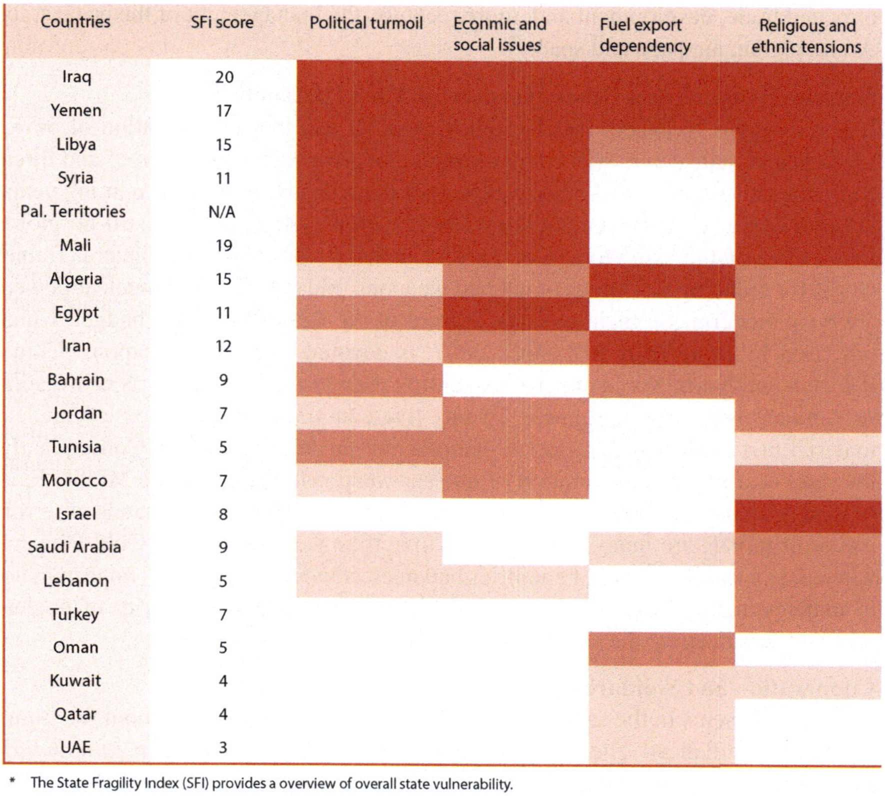

Fig. 5.4 Vulnerability of countries in the Middle East and North Africa. (Source: HCSS 2014: 198)

Fighting drags on, levels of violent crime remain high and development is stagnating. The result is labour migration and streams of refugees. Various forms of violence are also increasingly interconnected in these countries.

Local political movements, for example, are financed with money from criminal activities or by international terrorist movements that join them. Local political grievances, social and economic deprivation and increasing drought brought on by climate change increase the potential for more violent conflicts, terrorism and refugee crises.

Asia and South and Central America

The further growth of the economic and political importance of countries like China, Indonesia, India, Brazil and Mexico will probably translate into greater tensions in trade policy, rivalry over access to raw materials and transport routes and expansion of their military potential. Pressure will also increase on the international architecture constructed under the auspices of the UN. On a positive note, that architecture now extends to practically every social domain (domains that are in fact also covered by other international organisations): peace and security, human rights, economy and trade, development and, more recently, the management of the oceans, the Arctic region, airspace and space.[46]

Nuclear Weapons and Other Weapons of Mass Destruction
It is reasonable to expect that these power shifts and the fragmentation of power towards non-state actors will be manifested in a growing proliferation of and threat from nuclear weapons and other weapons of mass destruction in the coming years. Despite earlier US-led attempts to establish such a system, there is still no global monitoring of the safe storage of nuclear materials or robust multilateral frameworks for reducing stockpiles of nuclear weapons. Meanwhile, most nuclear powers are busy modernising their arsenals. A third of the Russian defence budget, which has risen by more than 50% since 2007, is devoted to nuclear weapons. China, Pakistan and North Korea are also expanding their arsenals, and the US is investing in a modernisation programme.[47] Twenty-five countries currently possess nuclear material that could also be used to manufacture nuclear weapons.[48] Consequently, the risks of a further proliferation of nuclear weapons to states in the Middle East, of misunderstandings and accidents and of weapons falling into the hands of terrorists or criminals are many times greater than they were during the Cold War era, when only a small number of countries had nuclear weapons. It is also important not to underestimate the issues surrounding the maintenance and security of nuclear weapons and keeping the security systems up to date.

Cooperation and Solidarity with Alliances
The developments in the security environment also raise questions about the future solidarity of alliances. The Clingendael Monitor 2016, *Grootmachten en mondiale stabiliteit* [Great powers and global stability] (which formed part of the Dutch government's Strategic Monitor together with the HCSS 2016 report *The Wheel of Fortune* referred to below), focused mainly on forms of cooperation and conflict between the great powers (including the EU). The most important conclusions were: cooperation and conflict are not mutually exclusive; there are many forms of ad-hoc cooperation (not involving the forming of blocs); the US-China relationship will grow in importance but its nature will change (specifically, it will be based less on trust) and will consequently be less stable than the US-EU relationship; relations between the great powers will become less predictable; and the international order will become less liberal. In its Strategic Monitor 2016 (*The Wheel of Fortune*), the HCSS analysed patterns of cooperation and conflict in the international relations between both state and non-state actors. Important observations made in the report were: forms of cooperation are diminishing; the major powers are becoming more assertive in pursuing their own direct interests; the number of major conflicts with large numbers of victims is increasing; classical inter-state crises have returned. However, the negative developments were balanced by some positive trends/sources of security (people worldwide are better educated, better fed, healthier and more tolerant).

The conclusion could be that the security environment in which the Netherlands and Europe find themselves has changed, but the possibilities for cooperation with other countries are also evolving. The stability of the relationships within the alliances will become less predictable with the United Kingdom's exit from the EU and with the course being taken by Trump's presidency in the United States. Concurrent conflicts with countries with which we have trade relations are likely to become more rather than less frequent, while the diversity of partners with whom action is taken in response to a threat or a conflict will increase, as will the forms of partnership. In that environment, the Netherlands' security policy (both national and international) needs to be reviewed – and the position of the armed forces in particular needs to be reassessed.

5.6 Changes in Warfare

It is impossible to say with certainty what future conflicts and what form of deployment the armed forces need to prepare for – the world is constantly changing. However, a number of dominant trends can be identified.

Hybrid warfare is receiving far greater attention as a result of Russia's actions in Crimea and in East Ukraine, where the Russian approach was characterised by the denial of military involvement (such as the so-called 'proxy war') and the use of non-military means (including propaganda, destabilisation of the public debate by distributing fake news and damaging the reputation of individuals and authorities, cyber warfare – including hacking and the use of 'trolls' on social media – and economic pressure).[49]

Hybrid warfare is the collective name for warfare involving the use of a range of military and non-military instruments (see Fig. 5.5). It is not an entirely new form of warfare. It offers states – and non-state actors – various possibilities to exert influence in the current security environment without engaging in large-scale military confrontations. Although large-scale military conflicts cannot be ruled out, hybrid forms of conflict and warfare appear more likely. The use of non-traditional instruments – not necessarily targeted at an enemy's military strength – can have a seriously disruptive effect and fundamentally change the traditional points of departure for operations.[50]

Warfare in urban environments is another scenario for future deployment of the armed forces. This is a logical consequence of the forecast that in 20 years' time the majority of the world's population will be living in 'megacities'. That will impose specific demands on military action, based on the use of technologically advanced tactical capabilities. Another dominant trend concerns the expectation that armed forces will be called on more frequently to tackle the effects of natural disasters, in their own country and elsewhere, as a consequence of climate change.

Technological developments will also influence the operational environment and the deployment of the armed forces – (see Fig. 5.5).[51] They include advances in communication (resulting in steadily greater integration, provided systems are compatible), observation, precision (targeted use), autonomy and automation (of weapons systems). These developments will probably benefit armed forces, but at the same time – in the hands of state and non-state enemies – will inevitably make

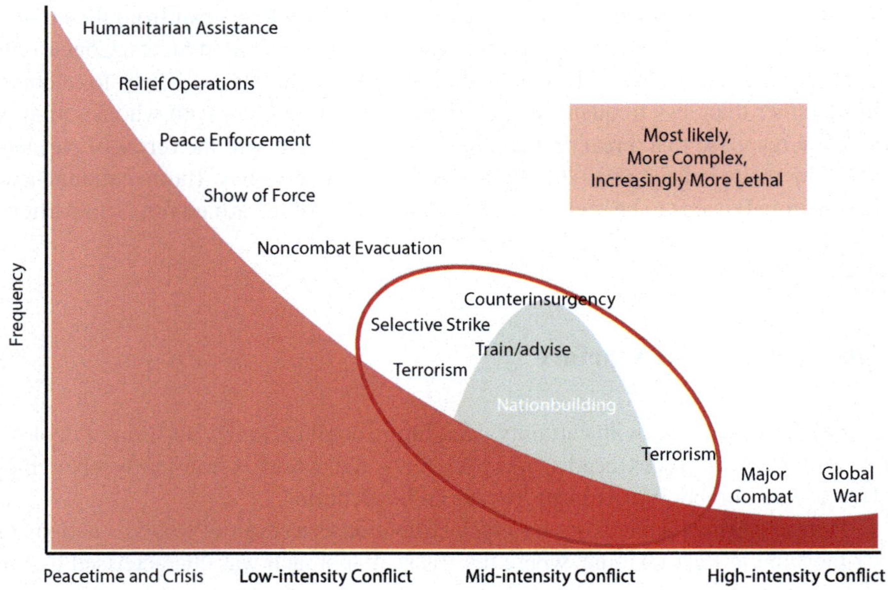

Fig. 5.5 Differences in involvement in conflicts and warfare. (Source: Hoffman 2009)

societies, and particularly their critical infrastructure, more vulnerable. That infrastructure could also involve (civil) capabilities in space, in the form of satellites. Achieving military superiority will be more challenging and more expensive because of the accelerating pace at which technologies are being developed and disseminated. Electronic warfare is an example of this. Effective access to the information domain – by gathering, analysing and using data (Big Data) (see the WRR's report *Big data in een vrije en veilige samenleving* [*Big Data in a free and secure society*] (2016)) – will be a critical success factor in military operations. Embracing these technological innovations is vital for future-proofing the armed forces, whether as a deterrent to (potential) enemies, for gaining and retaining the upper hand in (armed) conflicts or minimising the country's own vulnerability.

Whether this will lead to a major military upheaval, as was assumed in the 1990s, remains to be seen. With the shift from large-scale conventional (classical) action to mainly irregular and asymmetric warfare, the assumption that quick and resounding victories would come within reach has, in any case, not come about. In addition to the impact of technological developments, the importance of social development and the need to clearly understand the security environment are now recognised far more clearly than they used to be. Above all, the capacity to act in a complex security environment calls for knowledge and understanding of local conditions and the underlying causes of conflicts, both in the preparations and during the actual deployment.

5.7 Significance for the Main Tasks of Defence Policy

What do these developments imply for defence policy? That question can be answered on the basis of the three main tasks of defence policy:

1. protection of national and allied territory, including the Caribbean parts of the Kingdom;
2. promotion of the international rule of law and stability;
3. support for civil authorities in national law enforcement, providing disaster relief and humanitarian aid, both nationally and internationally.

What are the implications of the changed security environment for the performance of these three tasks? Political choices determine security policy. The deployment of the armed forces is not fixed, but is a consequence of political decisions. Nevertheless, defence policy rests on a number of constants that determine in part the relevance of the armed forces for security policy.

In the first place, the policy is not determined solely by the Netherlands' own choices and interests. Allies and partners expect contributions from the Netherlands. They also count on the smaller countries.[52] Secondly, the Netherlands has indicated that it wants to retain a high-class expeditionary force in the future in accordance with the principles laid down in *In het belang van Nederland* [In the interest of the Netherlands] (2013). This was the reason why a sum of approximately 0.5 billion euros was added to the defence budget in the 2014 Budget.

That political choice reflects the continuity of the policy of deploying the armed forces as an instrument for making relevant Dutch contributions to stabilisation operations that has been in place since the policy document on defence priorities *Een andere wereld, een andere defensie* [A different world, a different defence] in 1993. This choice is a perfect fit in the current era, with its heavy emphasis on the effects of instability in the regions around Europe for the country's own security. The defence of the territory of the country and its allies has 'reappeared' as a meaningful task of the armed forces. At the same time, this defence task will primarily involve making a relevant contribution to the defence of the alliance, since there is no question of a large-scale conventional threat against Dutch territory.

The Netherlands' participation in the US-led coalition against Da'esh illustrates the second constant in Dutch security policy: the contribution to intensive peace-enforcing operations under American leadership, under the auspices of NATO or otherwise. In light of the Russian aggression in Eastern Europe, the relationship with the US carries even greater weight than before 2014. The consideration that Dutch involvement in intensive peace-enforcing operations opens doors (and keeps them open) in Washington – and in the major European capitals – is as applicable as ever today. It creates possibilities for the Netherlands to exert influence.[53]

The possibility of contributing to both stabilisation and intensive peace-enforcing operations is also a reflection of the political spectrum in the Netherlands, with Europeanists and Atlanticists (both of whom are in fact to be found in all of the political parties) and their associated preferences for operations higher or lower in the spectrum of the use of force.[54] Accordingly, alternating coalitions have a menu of options, whereby parties can 'concede' missions to one another. For example, although there was broad political support for Dutch involvement in Iraq, the political debate about whether the mission could be extended to Syria underlined the fact that legitimacy and public support cannot be taken for granted. Politicians and the public are not *a priori* opposed to the use of force, but the aversion to genuine 'warfare' is deeply rooted in the Netherlands.[55]

The Assessment Framework for the deployment of the Dutch armed forces for crisis management operations illustrates the circumspection with which politicians operate. The political risks are not insignificant, as demonstrated by the fall of the fourth Balkenende government in 2010. The government and parliament therefore hold intensive consultations about Dutch contributions to crisis management operations, before the decision is made, during the operation and, in the last few years, on occasion after a mission has ended. In addition to the regular post-mission evaluation, the government can also decide to re-evaluate a mission 5 years after it has ended.[56] Strict demands are also made on the method of operation and the composition of Dutch military contributions. Examples are protecting the civil population or guaranteeing the necessary protection of soldiers sent out on missions.

5.7.1 Significance for the First Main Task

The changing security environment and the demands made on the armed forces give a specific interpretation and significance to the constitutional tasks. Treaty obligations mean that the first main task, the defence and protection of the interests of the Kingdom, naturally including the Caribbean parts of the Kingdom, has to be seen as including the defence of allied territory as well as the territory of the country itself. This has become far more relevant with the withdrawal of Russia as a cooperative security partner. Under the Readiness Action Plan, the mutual assistance clause, laid down in Article 5 of the NATO treaty, will be reinforced in the coming years in the form of more exercises, a larger presence with rotating units, aerial reconnaissance and enforcement along the Alliance's eastern border (see the NATO Wales Summit Declaration, 5 September 2014). These measures will be supplemented with more and heavier firepower (decision at the Warsaw Summit in 2016).

The greater instability in Eastern Europe has again drawn attention to the importance of the 'traditional' conventional capabilities that are required to form a credible deterrent.[57] In that context, NATO attaches great importance to Dutch military capacity such as F-35 fighter planes and submarines, as well as the Netherlands' contribution to ballistic missile defence with air defence frigates. That does not signify a return to the Cold War and the static defence strategy of that period, however. Military action today calls for a dynamic approach with mobile and rapidly deployable units, spearheaded by the Very High Readiness Joint Task Force (VJTF) from 2016.

NATO is prepared for a variety of scenarios, with special attention – because of Russia's actions in Crimea and East Ukraine and vis-à-vis the Baltic states – for hybrid warfare. That calls for specific capabilities in the field of cyber warfare, special units and capacity for intelligence gathering and analysis. Hybrid warfare is also receiving attention at EU level, where the emphasis is on the combined use of every available instrument, civil and military, in close cooperation with NATO.

Very important in this regard is the debate about what is known as A2/AD: Anti Access and Area Denial. With A2/AD weapons systems such as ballistic missiles or by disrupting communication, parties can be denied operational access to a region, both on land and at sea. Russia (near Kaliningrad and during the intervention in Syria) and China (the southern and eastern China Sea) have both created such areas. This not only prevents NATO from taking action during conflict situations, but also its ability to protect shipping routes, thereby also impairing flow security.

The following issues and the tasks derived from them require special attention in relation to this first main task:

Alliances with NATO and EU

The annexation of Crimea, Russia's intervention in Ukraine and various Russian statements, written and spoken, clearly illustrate the pressure on the international legal order in Europe. Deterrence and collective defence have assumed renewed significance. They demand an increase in NATO's state of readiness and compliance with the agreements made by the NATO countries in Wales (September 2014) and Warsaw (July 2016) to increase their defence efforts.

Estonia, Latvia and Lithuania, former Soviet Republics but now member states of NATO, are vulnerable and their defence requires special attention.[58] The Netherlands is playing its part in their protection, for example through its participation in the Very High Readiness Joint Task Force (VJTF) and Baltic Air Policing. This brings with it additional financial costs, in the short term to strengthen the sustainability of the Dutch armed forces and in the longer term to strengthen Europe's military capacity to act. Contributions to solving Europe's military shortfalls will not necessarily always comprise additional investments in personnel and materiel. More intensive cooperation is also an option, an area in which the Netherlands already has a good track record.[59]

Protection of Vital Infrastructure

The Netherlands and its inhabitants function by virtue of the country's vital infrastructure. The government and business community therefore work closely together to protect that vital infrastructure against disruption caused by disasters, attacks, technical faults and sabotage. In 2014, a new assessment of what constitutes the society's vital infrastructure was made on the basis of the economic, physical and social impact and cascade consequences, with a view to ensuring that instruments and scarce resources intended to enhance their resilience are employed as efficiently and effectively as possible. In that context, in the coming period the ministries of Justice and Security and Defence will continue the Enhancing Civil-Military Cooperation (VCMS) programme, thus creating a permanent partnership for crisis management between civil actors, such as the police, and the armed forces.

Counterterrorism

In addition to measures already taken, fighting jihadism, both in the Netherlands and at the source, is a priority. In the Netherlands, the NCTV – which is part of the Ministry of Justice and Security – is responsible for implementing the Comprehensive Action Programme to Combat Jihadism, which was launched in August 2014. The EU has also formulated a counterterrorism strategy. Focal points of that strategy are pursuing jihadist fighters and cutting off financing of terrorism and seeking cooperation with strategic partners, particularly in the Gulf region. With initiatives at bilateral, EU and UN level, the Netherlands focuses on eliminating breeding grounds for terrorism. Through the International Institute for Justice and the Rule of Law (IIJ), the Netherlands supports the process of judicial capacity building in North Africa and the Middle East. The intelligence and security services (the NCTV, AIVD and MIVD) play a crucial role in fighting terrorism and received additional funding for those activities in February 2015.

Flow Security: Energy, Trade, Raw Materials and Cyberspace

The world has become deeply interconnected. The Netherlands has one of the world's most open economies. Security is no longer location-bound, but depends to a large extent on unimpeded trade flows, uninterrupted availability of energy and raw materials and an open, free and secure Internet. The Netherlands must continue to actively campaign for the adoption of balanced rules and standards for global trade, which also benefit developing countries, in the World Trade Organisation (WTO) and through specific agreements such as the Transatlantic Trade and Investment Partnership. On the question of energy, the Netherlands should lobby in the EU for a reduction of the EU's dependence on Russia for gas.[60]

Cyber security is crucial for the Netherlands' prosperity and security (see also WRR (2015) *De publieke kern van het internet* [The public core of the Internet]). In the last few years, the Netherlands has taken steps to strengthen the country's resilience against state and non-state criminal activities with the formulation of a National Cyber Security Strategy (2011, 2013) and the establishment of the National Cyber Security Centre. The government, the business community, institutions and citizens will have to continue these efforts. In addition to defensive cyber capabilities (protection of their own networks, systems and information), the armed forces will also have to strengthen their offensive cyber capabilities (with digital instruments designed to disrupt or disable the enemy's actions). At international level, the Netherlands can promote international cooperation and the formulation of international rules. The organisation of the Global Conference on Cyberspace in April 2015 reflects the country's ambition of playing an active role in placing these issues on the agenda.

The Dutch armed forces also safeguard flow security with the deployment of the navy to combat piracy and to protect shipping routes (in addition to private security). This is a domain where the A2/AD phenomenon discussed above, which might make such operations impossible, could affect the Netherlands and in particular its trade interests and distribution channels. Innovation and resolve, but particularly training and exercises in dealing with these new phenomena in conflict situations, are required.

5.7.2 Significance for the Second Main Task

As the previous chapters have shown, particularly the second main task of the armed forces, the maintenance and promotion of the international legal order, has been called into question. Nevertheless, that task will remain as relevant as ever for the time being – in its own right and as an extension of the national interest (WRR (2010b) *Aan het buitenland gehecht* [Attached to the World]). Although large-scale interventions and the formation of the nation state have become less likely as a result of negative experiences in Iraq and Afghanistan, the greater instability in North Africa and other parts of Africa and the Middle East calls for active involvement, particularly on the part of Europe. Especially in Africa there will be a great need for stabilisation and peacekeeping missions, which should be based on an integrated approach to the complex and intransigent intra-state conflicts in the region.[61] The Netherlands is likely to be called on repeatedly in that context, in which case the armed forces could be deployed in various ways as part of an integrated Dutch contribution. This might involve peacekeeping operations with a broad mandate, including the possibility of using force, usually under the auspices of the UN. MINUSMA is the most recent example of such a mission.

The armed forces could also be regularly deployed for smaller-scale missions to train and advise the armed forces and the police in fragile states. The EU has been a frontrunner in this type of operation in the last decade and is also expected to develop security-related activities outside the traditional domain of the Common Foreign and Security Policy, for example in the context of the European Neighbourhood Programme, the Mediterranean Union and the Khartoum Process.

These are initiatives in which the armed forces are deployed in a preventive role in the pre-conflict or post-conflict phase. A specific consideration in relation to the deployment of the armed forces in the context of their second main task is the desired duration of Dutch involvement in reducing instability. The government recognises this and refers to the long-term effort required to tackle the underlying causes of instability. In terms of the specific military capabilities that are needed, the MINUSMA mission in Mali can serve as an example. The main requirements are intelligence (gathering and analysis), special units, firepower (in the context of the desired escalation dominance, in this case with attack helicopters) and mobility (transport helicopters). The use of units operating in remote areas and under difficult circumstances (in terms of terrain, climate and accessibility) also makes major demands on logistics and support.

The following issues and the tasks derived from them require special attention in relation to this second main task:

Prevention

Prevention is better than cure, particularly when it comes to crises and conflicts. The International Security Strategy refers to the importance of early warning and prompt action. This depends mainly on timely access to adequate information. In the meantime, additional information becomes available via the intelligence and security services, NGOs supported by the Netherlands and international organisations. However, the capacity to respond to early warnings – independently; with or via state and non-state partners; or via international organisations – is still very limited. The capacity to identify warnings in time also needs to be improved at international level. The EU and NATO were able to formulate responses to the Ukraine crisis, but, like the individual member states, ignored or underestimated the early warnings. Measures to strengthen the capacity for preventive action must therefore also include putting threats that have been identified on the agendas of international organisations and with the member states. That calls for (additional) mechanisms, starting with the development of preventive policy detailing the specific possibilities for national and multilateral measures to prevent conflicts.

Fragility: Tackling Instability at the Source

Fragility is one of the major causes of instability. The Netherlands has implemented 3D policy and fragile states policy and promoted an integrated approach in a series of crisis management operations, in combination with diplomatic and aid efforts. In the process, the relevant ministries, NGOs, the business community and knowledge institutes have acquired relevant knowledge and expertise. For example, networks have been created in which the actors concerned work closely together in preparing and carrying out operations. *De Leidraad Geïntegreerde Benadering* [Guidelines for the Operationalisation of the Integrated Approach] (2014), in which cooperation is identified as the critical success factor, reflects that trend. However, the integrated approach was developed mainly as an executive instrument in the context of specific operations.

The experiences with the fragile states policy (see Chap. 3) should be translated into action perspectives, however difficult that may be. After all, the fragile states policy dates from 2008 and *De Leidraad Geïntegreerde Benadering* (2014) focuses mainly on collaboration between the various actors, but there is no strategic long-term plan that takes account of the underlying causes of instability and the possibilities of preventive action. An integrated prevention and stabilisation unit within the Ministry of Foreign Affairs (including Foreign Trade and Development Cooperation), with the participation of all the relevant ministries, could develop such an approach and then coordinate all of the activities dedicated to preventing and reducing instability in fragile states and regions.

Even with the further development of the integrated approach, it will still be necessary to anticipate the need for military interventions in the future. Wherever genocide, serious human rights violations or other humanitarian emergencies occur, military intervention must, if necessary, still be an option in light of the principle of the 'responsibility to protect' (the responsibility to protect the populations of other countries from mass atrocities) and invoking human security.

Support of States in Eastern Europe, the Middle East and North Africa

The Netherlands should actively concern itself with the position of countries in the 'ring of instability' in Africa and the Middle East that have not (yet) descended into chaos and conflict. Cooperation with these countries, particularly Algeria, Egypt,

Jordan, Lebanon, Morocco and Tunisia, is required. These countries are struggling under difficult circumstances and are important players in the region.

The Netherlands should review the possibilities of intensifying the cooperation with these countries, both bilaterally and via the cooperation programmes of the EU and NATO. In Eastern Europe, where Russia is a major source of instability, various countries are facing similar circumstances, in particular countries that are not members of NATO or the EU, but would like to join them. The security and stability of these countries is threatened by the disruptive activities of Russia. Balkan states such Kosovo, Bosnia and Macedonia are also far from stable and there lies an important overlap with the European Neighbourhood Policy that has been developed since 2004.

Promoting the Rule of Law, the Legal Order and Human Rights

Violations of the international legal order, such as Russia's recent annexation of Crimea, must not go unpunished. Countries must be held accountable for their actions. The same applies with regard to respect for human rights. After all, in every country that is currently experiencing instability, the rights and freedoms of citizens, regardless of their religion or ethnicity, are not adequately protected. Personal security demands a political and civil order that protects everyone's rights and freedoms, not just those of certain groups. The Netherlands should continue to press as hard as ever to promote the rule of law, the legal order and human rights, in the firm conviction that a stable world order is only possible if the rights and freedoms of citizens are protected worldwide, including those of citizens of fragile states. But those efforts will have to be in proportion to what can actually be accomplished by exerting influence in the geopolitical arena. The question of how values can be conveyed cannot be seen separately from the question of what type of 'power' is needed to actually give effect to those values. Promoting the international legal order will therefore remain a main task of the armed forces.

Investing in Global Institutions, Particularly the UN

Global institutions such as the UN, the World Trade Organisation (WTO) and the World Bank are under pressure, partly because of the diminishing interest among emerging countries in participating in institutions in which they do not feel properly represented. This development is also the result of regionalisation, a process in which groups of like-minded countries make agreements among themselves. The Netherlands is rightly concerned about this trend, since the international (legal) order depends on these institutions. They are moreover the primary vehicles for reaching agreement on global issues such as climate change. The Netherlands should continue to lobby forcefully for an international order based on legal rules and agreements and should be willing to reach compromises with countries that demand a greater role. The modernisation of institutions extends beyond a redistribution of influence. Reforms are also necessary.

5.7.3 Significance for the Third Main Task

Because of the growing interconnectedness of internal and external security, the third main task of the armed forces, providing support for civil authorities in upholding the law and providing disaster relief and humanitarian relief, both nationally and internationally, will grow in importance. After all, the armed forces are directly involved in preventing and reducing external threats ensuing from the greater instability in the region around Europe and the greater vulnerability of the Netherlands in globalised international relations. But their role also includes coordinating and supplying part of the coastguard's capacity and providing unique capabilities in explosive clearance, aerial reconnaissance, firefighting and special assistance. There will be growing demand for these capabilities.

The following areas of attention and ensuing tasks require special attention in relation to this third main task:

Assistance in Controlling the Borders of the Netherlands and Europe
Border control is an important aspect of the armed forces' role as a permanent security partner of the civil authorities. With the emergence of border security as a new priority, the third main task will only become more explicitly devoted to this aspect. In addition to the operational issues, the support of border security ensuing from defence policy should also be explicitly addressed in the policy on Europe. With border security increasingly justifying European cooperation, defence policy will become an intrinsic component of the thinking about future European cooperation.

Civil-Military Cooperation in Response to Disasters or Calamities
Another important role for the armed forces in support of the civil authorities is to be on call to prevent or mitigate social disruption in the event of natural disasters or human-induced calamities in the context of the Intensifying Civil-Military Cooperation (ICMS) programme. Some of the necessary military capabilities are provided by specific branches of the armed forces, in particular the Royal Netherlands Marechaussee (which is statutorily responsible for border policing, guarding and securing important objects and individuals and performing international and military policing tasks), and some are regular military capacities, such as F-16 fighter planes. A substantial proportion of the armed forces can be called on for the purposes of the ICMS.

The changes in Europe's immediate security environment have major consequences for defence policy and the performance by the armed forces of their tasks. The focus since 1993 on the second main task, maintaining and promoting the international legal order, needs to be recalibrated given the growing demands being made on the armed forces for the country's own security, both in the context of their defence task and in support of the civil authorities. Furthermore, in relation to the first two main tasks there is a need to take greater account than in the past of a simultaneous, additional call on military capacities in the event of a deteriorating security environment in Eastern Europe or a calamity in the Netherlands that demands an immediate response. The significance for the third main task is that these tasks are not clearly distinguishable in territorial terms but largely overlap, and that national and international are increasingly connected with one another. Examples are growing threats of terrorism as a result of the Netherlands' interventions in other countries and the continuing instability in the region around Europe. Tensions in the regions often have historical, cultural and economic connections with Europe, European policy and European interventions – or lack of them – so that external security is interconnected with internal security.

Defence policy must therefore be seen explicitly in the context of the strong mutual dependencies between the national, regional and global levels. The main tasks in themselves remain as important as ever, but their significance is changing because they have to be consistently understood in the context of the interconnectedness of internal and external security. The altered significance of the main tasks as outlined in this chapter provides a template for setting substantive priorities in security policy and the ensuing choices with regard to the resources of the armed forces discussed in the next chapter.

Endnotes

1 Article 122 of the 1814 Constitution; Article 204 of the 1815 Constitution was almost identical.

2 The concept of defence by means of inundation was used once again in the early 1950s, although aircraft and missiles had greatly diminished its value, with the construction of the IJssellinie [IJssel Line] as a water-based line of defence further to the east. See Reijer, E.C. The (1997) *De IJssellinie 1950–1968*, Zeist, Rijksdienst voor de Monumentenzorg (Heritage Management Agency) and Waanders Uitgevers, Zwolle.

3 Weidenhaus, G. (2015: 199).

4 Rosenberg, E.S. (2012).

5 Judt, T. (2005: 242).

6 Kissinger, H.A. (1957/1969: 125).

7 Rood, J. (2010).

8 Judt, J. (2005: 247 ff.).

9 Winkler, H.A. (2015: 189).

10 Hirsch Ballin, E.M.H. (2014: 138–139).

11 Final report of Future Policy Survey 2010 – A new foundation for the Netherlands Armed Forces: 276; 303, annex to Parliamentary Documents II 2009–2010, 31,243, no. 17.

12 Parliamentary Documents II 2010–2011, 32,500 X, no. 41.

13 Winkler, H.A. (2015).

14 Winkler, H.A. (2015: 357 ff.).

15 Korteweg, A.R. (2011: 227); Wijk, R. de (2014: 62).

16 Schoenmaker, B. (2012: 6–7); Graaf, B. de (2012d: 52); Korteweg, A.R. (2011: 217).

17 Abels, P.H.A.M. & Willemse, R. (2004) outline how the declining significance of the communist threat due to the West's growing technological supremacy and the potential threat of the spread of Islamic fundamentalist ideology under the influence of the Iranian revolution and the mujahedeen's victory over the Russian army in Afghanistan in 1979 was already anticipated in the 1980s.

18 Wijk, R. de & Toxopeus, R. (2005: 423).

19 Korteweg, A.R. (2011: 258).

20 Parliamentary Documents II 2013–2014, 29,521, dated 22 January 2014.

21 Biscop, S. (2015).

22 Voorhoeve, J.J.C. (2010).

23 AIV. (2012: 12–13); Van Staden 2013: 64)

24 Homan, C. (2006).

25 Homan, C. (2007: 63).

26 Homan, C. (2007).

27 AIV. (2012: 16).

28 AIV. (2015).

29 Parliamentary Documents II 2014–2015, 21,501–20, no. 997 dated 26 June 2015.

30 CEPS. (2015: 5).

31 Parliamentary Documents II 2014–2015, 21,501–20, no. 997, 26 June 2015.

32 Palm, T. (2016).

33 Drent, M. et al. (2016); Bakker, A. et al. (2016).

34 Biscop, S. (2015).

35 Germany took the lead in the Ukraine crisis and the negotiations with Russia, while France focused on the Sahel and Italy was in the vanguard in the Libyan crisis. As the AIV observed, such a division of responsibilities can work well, but in time can also cause alienation that stands in the way of joint decision-making. Other member states, such as Poland, might feel excluded or feel prejudiced by the economic sanctions because of their greater economic dependence on Russia. Conflicts of interest of this nature will impose limitations on the strength of the NATO alliance (AIV 2015).

36 Odgaard, L. (2014); Drent, M., Landman, L. & Zandee, D. (2014); Parliamentary Documents II 2014–2015, 28,676, no. 225, 12 June 2015.

37 Parliamentary Documents II 1993–1994, 23,790, no. 3, p. 9.

38 M.G. Fischer and D. Thym, Commentaar op artikel 42 VEU [Commentary on Article 42 of the Treaty on European Union], no. 46, in: H.-J. Blanke & St. Mangiameli (eds.), *The Treaty on European Union (TEU)*, Berlin/Heidelberg: Springer 2013, p. 1226.

39 Venezuela's constitution contains a provision which allows other territories to join Venezuela by means of a freely adopted decision of its inhabitants: Artículo 14. La ley establecerá un régimen jurídico especial para aquellos territorios que

por libre determinación the sus habitantes y con aceptación the la Asamblea Nacional, se incorporen al the la República.

40 Griffith, I.L. (1996: 19).

41 Griffith, I.L. (1996: 33).

42 Prevost, G. & Vanden, H.E. (2014: 84).

43 Freedman, L. (2014: 18); Persson, G. & Pallin, C.V. (2014: 30–31).

44 Freedman, L. (2014: 8); Granholm et al. 2014; Rood et al. 2015: 8.

45 Hoffman, F.G. (2009); Freedman, L. (2014: 20); Granholm, N. & Malminen, J. (2014: 10).

46 SWP & GMF (2013:12); Missiroli, A. et al. (2014); Anthony, I. (2014: 5); HCSS (2014).

47 *Economist* 2015, 7 March.

48 Missiroli, A. et al. (2014: 34).

49 CSIS (2016); see also House of Commons Defence Committee 2016.

50 Haaster, J. van & Roorda, M. (2016).

51 Ministry of Defence (2014); U.S. Marine Corps (2015).

52 Archer, C. Bailes, A. & Wivel, A. (2014); Lindley-French, J. & Tjepkema, A. (2010).

53 Wijk, R. de (2014); Korteweg, A.R. (2011); Lindley-French, J. & Tjepkema, A. (2010).

54 Korteweg, A.R. (2011).

55 Lindley-French, J. & Tjepkema, A. (2010).

56 House of Representatives, 2011–2012, 29,521, no. 191.

57 Casteleijn, L. (2015).

58 Rand Corporation (2016).

59 McKinsey and Company (2010).

60 AIV. (2014).

61 Clingendael Strategic Monitor (2014).

Chapter 6
Strategic Strengthening of the Armed Forces

Ernst Hirsch Ballin, Huub Dijstelbloem, and Peter de Goede

6.1 Introduction

Against the background of the profound changes in the Netherlands' security environment and increased instability in the region around Europe and the challenges that need to be addressed by defence policy sketched in the previous chapter, the question is whether the armed forces are capable of adequately performing their tasks. This chapter investigates the state of the armed forces and defence policy and considers what the focus should be and what investments are needed. The discussion centres on the following questions:

- What is the current state of the armed forces? (Sect. 6.2)
- What is the status of defence policy? (Sect. 6.3)
- What investments are needed and where is strengthening required? (Sect. 6.4)
- How can the possibilities for European cooperation be used more effectively? (Sect. 6.5)
- What policy is needed to accomplish that? (Sect. 6.6).

6.2 The Current Armed Forces: Modernised, but with Tight Constraints

The answer to the question about the 'state' of the armed forces starts with the political choice made after the Cold War to make exporting stability a new spearhead of foreign and security policy. Expeditionarity, the deployment of the armed

E. Hirsch Ballin (✉) · H. Dijstelbloem · P. de Goede
WRR, Scientific Council for Government Policy, The Hague, Zuid-Holland, The Netherlands
e-mail: ballin@tilburguniversity.edu; h.o.dijstelbloem@uva.nl; peter.goede@minbzk.nl

E. Hirsch Ballin et al. (eds.), *Security in an Interconnected World*,
Research for Policy, https://doi.org/10.1007/978-3-030-37606-2_6

forces far from the Netherlands, became the point of departure for the structure of the armed forces. Conscription was suspended; the armed forces would in future consist of professional soldiers who could be deployed on missions and part-time reservists. Their materiel consisted, among other things, of a range of transport equipment including heavy Chinook transport helicopters, Landing Platform Docks (transport and command vessels) and capacity for tactical and strategic air transport. In addition to the modernised F-16 fighter planes, which could in future be refuelled in the air by KDC-10 tanker planes, and air-defence and command frigates and submarines, the Netherlands possessed valuable capabilities for operating at every level in the spectrum of the use of force. This meant that the armed forces could be deployed both for military interventions and combat operations and for stabilisation and peacekeeping missions. The distinction between these types of mission would blur over time, with the Dutch mission in Uruzgan as the tipping point. There was also investment in interoperability, an essential requirement for taking part in international operations, since there was no question of engaging in international operations independently. The Netherlands contributed to EU, NATO and UN operations and joined US-led coalitions.

With this transformation from the 1990s, in combination with continuous contributions to crisis management operations, the Dutch armed forces gained a wealth of experience and earned a good international reputation.[1] The Netherlands became a 'member of the A-team', the small group of countries with interoperable, high-class military capabilities that was a close military and political partner of the US within NATO or as part of a coalition.[2] In the Netherlands, the armed forces evolved into a structural partner of the civil authorities, for example through the key role played by the Royal Marechaussee in border control, the establishment of the coastguard with a coordinating role for the navy, and the intensification and strengthening of civil-military cooperation (CIMIC/Enhanced CIMIC) in the wake of the 9/11 attacks in 2001. The Air Mobile Brigade was also formed. The cooperation with the Ministry of Foreign Affairs and the directorates-general for foreign trade and international cooperation was intensified with the emergence of the 3D approach, in which the collaboration between Dutch military personnel, diplomats and aid workers in preparing and carrying out crisis management operations grew steadily closer. Beyond the national borders, the cooperation with Germany and Belgium assumed a structural character. The existing naval cooperation between the Netherlands and Belgium in BENESAM was further intensified.[3] The deployment of reservists was classified under the Total Force concept, TFC (see *Plan van aanpak uitvoering Total Force concept* [Action Plan for the implementation of the Total Force concept], 13 January 2017).

The changes and the austerity measures, in combination with active deployment, created evident problems for the armed forces, for example in terms of the physical and psychological strain on military personnel, the shortages of materiel for operations and exercises and in arranging the logistics and support for large numbers of troops. Nevertheless, the transformation of the Dutch armed forces into a high-class expeditionary force is generally regarded as a success. Evidence of this is provided by the significant contributions currently being made to crisis management operations, in particular the UN's MINUSMA operation in Mali, the US-led coalition against Da'esh and the EU-led mission to combat piracy around the Horn of Africa. Another

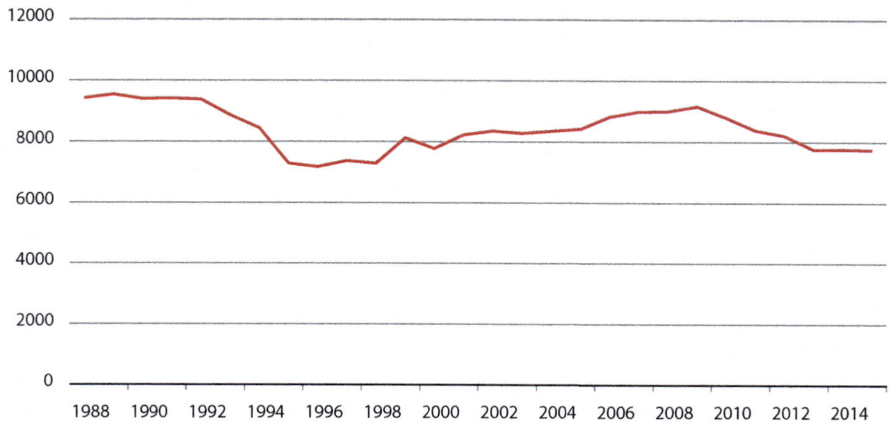

Fig. 6.1 Defence spending in 2014 (in euros). (Source: SIPRI Military Expenditure Database 1988–2015 (1 USD = 0.73 Euro, exchange rate on 30 June 2014))

example is the Dutch contribution to the measures taken by NATO to reassure the eastern allies, including participation in the Very High Readiness Joint Task Force (VJTF) and the air policing over the Baltic States.

The modernisation of the armed forces has been accompanied by cutbacks, initially imposed as a peace dividend and later because the targets in the Stability and Growth Pact were not being met. As a result, actual spending on the armed forces fell sharply between 1995 and 2013 (see Fig. 6.1). Expenditure declined by more than 25% in real terms. A number of capabilities were disposed of entirely, including the last two tank battalions (in 2011), maritime patrol aircraft, multiple launch rocket systems (MLRS) and air defence systems. The workforce was incrementally reduced by more than half. With these measures, the rapidly escalating operating and investment costs could also be absorbed.[4] Finally, the long-term deployment of military capacity outside the national territory and under severe conditions generated unforeseen costs, for example due to accelerated wear and tear on equipment,[5] some of which were covered from the Homogeneous Budget for International Cooperation (HGIS) and supplements to the defence budget. The mission in Afghanistan in 2008, for example, forced additional cutbacks. By spreading the pain as evenly as possible, the individual branches of the armed forces, the army, the air force and the navy, were able to retain high-class capabilities. Reductions in the numbers of systems, platforms and units were accepted, which can be traced back to a large extent to the long-accepted but now formally abandoned formula for the allocation of funds of 50%, 25% and 25% to the army, the air force and the navy, respectively.[6] The allocation of funds in 2016 is shown in Fig. 6.2.

There are various explanations for the cutbacks on the armed forces. The end of the Cold War marked the end of their exemption from austerity measures. From that time on, the Ministry of Defence had to compete with other ministries for funds. In combination with continuous unexpected shortfalls as a result of escalating operating and investment costs, the result was a dramatic reduction in the size and composition of the armed forces. The so-called 'differentiated approach', under which new

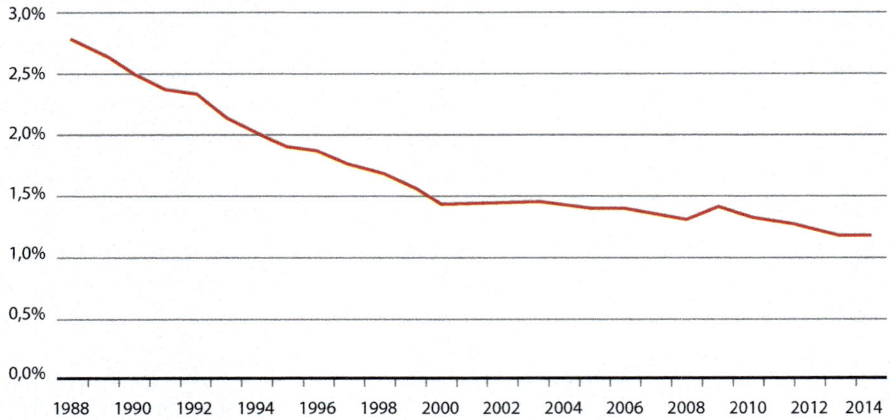

Fig. 6.2 Defence spending as % of GDP. (Source: SIPRI Military Expenditure Data 1988–2015)

investments in the armed forces were financed from spending cuts, also created the perception that there was always some fat that could be removed from the armed forces. In that context, Relus ter Beek, who was Minister of Defence from 1989 to 1994, compared the armed forces with a lemon. No matter how long you keep squeezing it, some juice will always escape ….[7] And indeed the armed forces could be 'squeezed' for a long time. The branches of the armed forces enjoyed a large degree of autonomy. That came to an end in 2003, but because of the differentiated approach the idea that there was always room for more cutbacks on the armed forces survived for a long time (see Fig. 6.1). An additional factor was that cutbacks were specifically used to steer the armed forces further in the direction of expeditionarity. The most important explanation, however, was the key position that the desire to make contributions to international crisis management operations occupied in defence policy and the associated level of ambition that had been formulated for the armed forces. That level of ambition could be lowered without major political consequences. Contributions could still be made; the security of the Netherlands was no longer threatened since the end of the Cold War (Table 6.1).

Table 6.1 Division of expenditure in 2016 and 2017 in millions of euros

	2017	2016
Navy	721,48	689,55
Army	1.251,34	1.137,98
Deployment	325,79	367,89
Air Force	681,68	633,80
Military Police	52,41	319,52
Defence Materiel Organization	830,34	743,89
Central Apparatus	1.580,73	1.595,55
Investments	1.640,54	1.446,20
General	95,72	100,14
Support Command	1.162,45	1.040,57
Secret Expenditure	5,39	5,35
Nominal and Unforeseen		153,48

The Netherlands therefore got its expeditionary force, but ultimately paid a heavy price in the form of far smaller armed forces with a significantly lower level of ambition. In the 1990s, the armed forces were still expected to be capable of participating in four crisis management operations; in 2013, the number was only one. The sustainability of the armed forces, their capacity to contribute to an operation for a lengthy period, also diminished greatly.

Because of a shrinking reservoir of units, capabilities and personnel, in combination with the repeated choice of sparing the operational units – combat strength – as far as possible, missions also started to impose a growing strain on the organisation. Recent letters to the House of Representatives indicate the seriousness of the resulting problems, which affect the army, the air force and the navy: on the issue of the state of readiness of materiel, see the letters of 9 October 2014[8] and of 26 May 2015.[9] There are shortages of spare parts, technical personnel and materiel for training. The minister refers in that context to the impact of these constraints on the capacity to carry out simultaneous operations. In its audit of the Ministry of Defence in 2014, the Netherlands Court of Audit was severe in its assessment: "At this moment the organisation is unable to retain sufficient materiel for training and education purposes in addition to the deployment for missions" and "There are tight constraints on the available resources for many of the tasks" (Netherlands Court of Audit, *Resultaten Verantwoordingsonderzoek 2014 Ministerie van Defensie* [Results of the audit of the Ministry of Defence 2014]).

6.3 Defence Policy: Ambitious, but Confronted with Shortages

Against the background of the growing instability around Europe and the current problems facing the armed forces, the question that arises is whether the current defence policy can put an end to the persistent struggle between tasks, ambitions and resources. As Fig. 6.1 shows, since the end of the Cold War successive governments have made deep cuts in defence spending. Relative to GDP, the budget has been more than halved (see Fig. 6.2). The same applies for many of the other allies.

Under NATO agreements every ally is obliged to spend at least 2% of its national income on defence. Many European countries have been failing to meet this target since the fall of the Berlin Wall (see Fig. 6.3). The Netherlands also fell below the threshold in the mid-1990s. In the most recent *Defence Planning Capability Review* for the Netherlands in March 2016, NATO concluded that a higher and more predictable defence budget is essential. The Netherlands currently spends 1.14% of the gross domestic product on defence, which is less than the average of 1.43% for the European NATO members and well below the NATO target of 2%. At the NATO summits in Wales (September 2014) and Warsaw (July 2016), the NATO member states that spend less than 2% of their GDP on defence committed themselves to endeavouring to raise their defence spending towards the target over the next 10 years. A

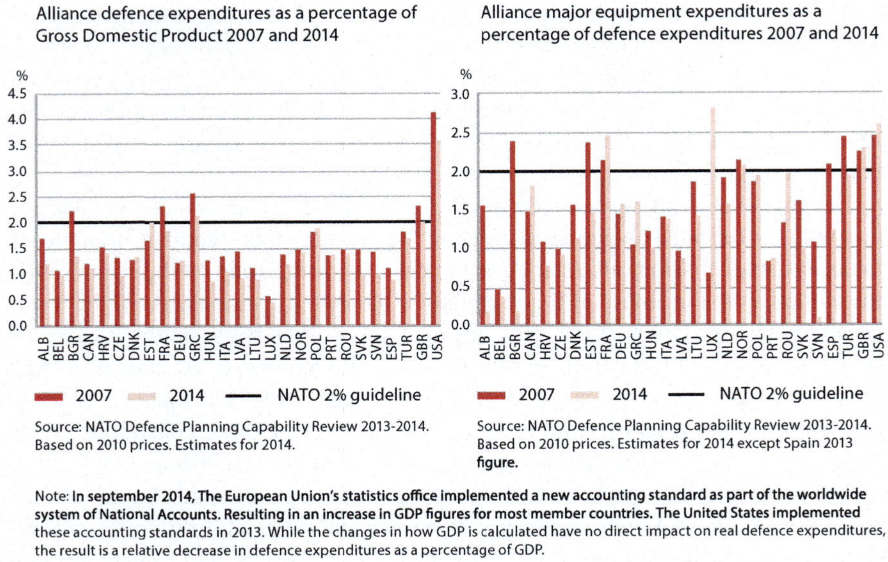

Fig. 6.3 European shortfalls. (Source: AIV 2015)

similar agreement was made in relation to the guideline of spending 20% of the total defence budget on investment. Agreements were also made to spend 2% of the defence budget on research and technology.

France and Germany are increasing their defence spending substantially in the coming years (by approximately 18% up to 2020–2021), but even then both countries will remain below the 2% target. A number of eastern NATO member states are also investing more. The United Kingdom has announced that it will maintain its defence spending at the desired level of 2% in the coming decade. The Baltic States (Estonia, Latvia, Lithuania) have increased their defence budgets.[10] Sweden and Finland, neither of them members of NATO, have also raised their defence budgets. In the Netherlands, the increase in the defence budget will stabilise defence spending at around 1.1% and, depending on the growth of GDP, possibly 1.2%.

Whether the anticipated availability of additional resources will be sufficient to achieve the desired strengthening of defence capabilities is debatable. Nor will the extensive military cooperation – bilateral, in clusters (groups) or in multinational structures with coordination by the EU and NATO – automatically eliminate the shortfalls. Cooperation also costs money because of the transaction costs. Even many large and medium-sized European countries simply lack the economies of scale and the financial resources required to provide the necessary capabilities on their own. A joint and coordinated approach to making up the European military shortfalls should therefore remain as much of a priority as ever even if defence spending stabilises or is raised slightly.

The annexation of Crimea in March 2014 shed an entirely new light on the decision to reduce the level of ambition for the armed forces at the end of 2013. NATO responded almost immediately with a series of measures designed to reassure the eastern allies. The Netherlands also contributed to those measures by deploying tanker aircraft, establishing a maritime presence and by promising to bring forward the promised deployment of fighter planes to help in protecting the air space over the Baltic States. It was perfectly clear to the government that the increased demands on the Dutch armed forces would not be merely temporary. Ahead of the NATO summit in Wales in September 2014, the NATO member states prepared a Readiness Action Plan designed to increase NATO's state of readiness. The deterioration in the security environment in Eastern Europe reawakened the debate about the shortcomings of the European NATO allies' defence efforts. The US exerted heavy pressure with a view to sending a powerful signal, to both Moscow and the eastern member states. The mood among politicians and the public had also changed radically.

The structural addition of 100 million euro to the defence budget in 2015, somewhat tentatively announced as a trend break in the Budget Memorandum, therefore came as no surprise, any more than the outcome of the General Political Debate (*Algemene Politieke Beschouwingen*), when the House of Representatives debates government policy following the presentation of the Budget. The SGP submitted a motion, which was also signed by the VVD, the CDA and ChristenUnie, calling for a statement by the government on the necessary level of ambition for the armed forces and the appropriate security strategy in the existing security environment.[11] This motion by the leader of the SGP, Kees van der Staaij, gained further momentum with the government's widely-supported decision to send six F-16 fighter planes to join the US-led coalition against Da'esh.[12] With contributions to Resolute Support, NATO's training mission in Afghanistan, and MINUSMA, the UN peacekeeping operation in Mali, the government once again made heavy demands on the Dutch armed forces. The pace of operations would accelerate even further with the anticipated contribution to NATO's Very High Readiness Joint Task Force (VJTF).

With the measures taken in 2014 and 2015 and the addition of 60 million euro to the International Security Budget, making a total of approximately 0.5 billion euro, important steps can be taken to resolve the most urgent problems facing the armed forces. But restoring capacity in terms of manpower and materiel will require more than the current budgetary impulses. Meanwhile, the multi-year perspective and the steps identified in it illustrate the deep-seated problems facing the Ministry of Defence, as is shown by Fig. 6.4. The planned strengthening of the supporting operational units underlines the fact that even with the measures that have been taken, there is insufficient balance in the armed forces. The reference to the replacement of essential capabilities leads one to assume that the planned investments cannot or not entirely be accommodated within the budget.

Significant efforts, financial and otherwise, will be needed to take the next steps in the envisaged multi-year perspective; the necessary funds have not (yet) been made available.

A more fundamental issue than the affordability of the existing armed forces and the armed forces as envisaged in the current defence planning concerns the future-proofing of defence policy as set out in *In het belang van Nederland* [In the interest of the Netherlands] (2013) and most recently in *Houvast in een onzekere wereld.*

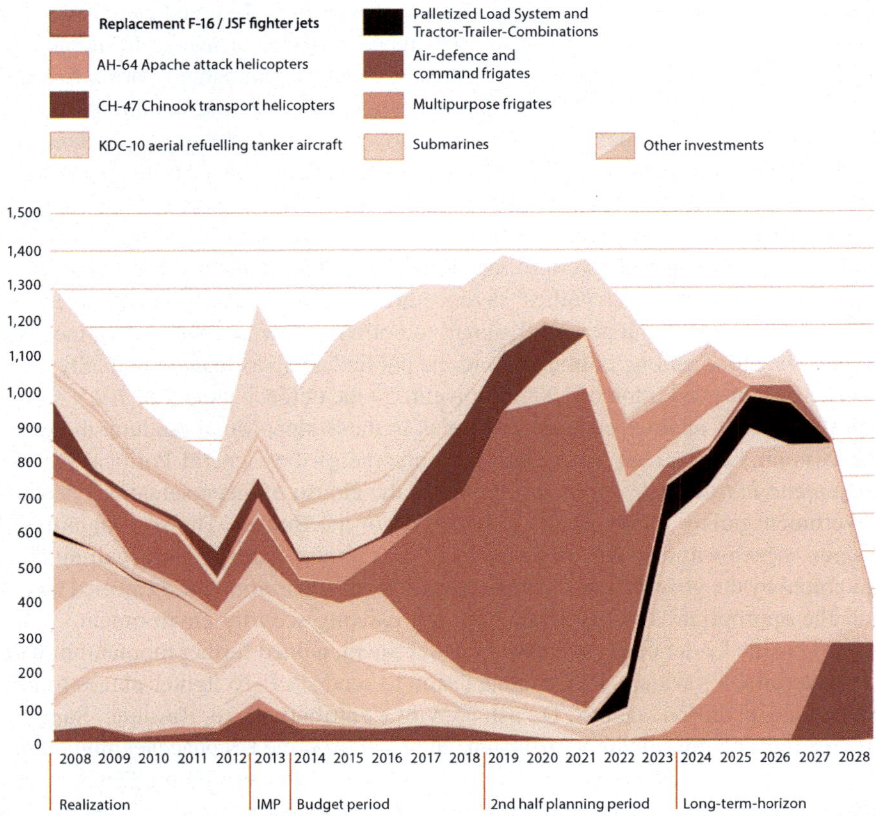

Fig. 6.4 A forward view of defence planning up until 2028. (Source: Netherlands Court of Audit)

Lijnen van ontwikkeling in het meerjarig perspectief voor een duurzaam gerede en snel inzetbare krijgsmacht [A grip in an uncertain world. Lines of development for a sustainably ready and rapidly deployable armed forces] (2017). In light of the profound changes in the Netherlands' security environment, it is time for a new reappraisal of defence policy.

The general points of departure of defence policy have not changed since the drafting of the *Defence Priority Review* (1993). The choice for versatility, for the retention of the widest possible range of capabilities, is based on the principle that a diffuse and unpredictable security environment calls for the widest possible range of options for deployment. Capabilities are not linked to specific threats, but to a series of (potential) risks. These points of departure need to be revised. A new policy and assessment framework for the size and composition of the armed forces is required. In Dutch defence policy, the questions 'what are we planning for?' and 'how much is enough?' are still answered on the basis of the level of ambition, the targets for deployability in relation to the armed forces' second main task and the desire for versatility and diversity in light of threats and risks to security – far from home. That approach is no longer sufficient, however. In view of the greater instability and insecurity surrounding Europe, the question is no longer which missions to participate in, but what is the Netherlands willing to do for its own security and that of its allies in light of specific threats in and around its own home in Europe? In other words, security is no longer a question of informal – higher or lower – ambition, but of necessity.

In the first place, versatility and diversity are no longer adequate as points of departure for the composition of the armed forces. The future remains unpredictable, but the threats are not diffuse, but specific. The Netherlands is confronted with direct threats – 'old' and 'new', around Europe and global – and constant demands will be made on the Dutch armed forces in that context. The degree of uncertainty about the future should not be exaggerated; the instability around Europe, its (possible) effects and the negative consequences of globalisation are palpable. Retaining, replacing or procuring capabilities should be matched as closely as possible to the existing and anticipated threats. This means the crucial question is: what range of capabilities can contribute most effectively to countering those threats now and in the future? Answering that question provides the most robust basis for making decisions about the future size and composition of the armed forces.

Secondly, the criteria for the choice of capabilities in the defence planning process in accordance with the current defence policy result in the retention of basic and (as far as possible) niche capabilities. That inevitably leads to the retention and replacement of existing main weapons systems. Consequently, there is, by definition, little scope for new capabilities. The room for innovation and change is further constrained by the (financial) relationships between the operational commands of the army, the air force and the navy and the sequential nature of the major investment programmes. After the purchase of the F-35 fighter planes, the replacement of the navy's main weapons systems is planned.

A look at the planning overview for the armed forces, assembled in the so-called 'template' in the defence budget, does indeed suggest that there will be investment in modernising and replacing main weapons systems, but scarcely any investment in innovation (see Fig. 6.5). The fate of MALE UAV, the unmanned aerial reconnaissance system, speaks volumes. The need to procure the system was recognised as early as 2006, but its purchase has again been deferred for 7 years in order to accommodate the investment in the financial planning. Accordingly, the Dutch contribution to reducing an important military shortfall in Europe remains just a proposal. The scope for investment in the cyber domain is also very limited in the present system when compared with the replacement investments. The same applies for other resources, such as special units to respond to a future in which military conflicts are most likely to take the form of hybrid warfare. At a time when innovations and experiments with new concepts are becoming increasingly important, the chance that the armed forces will in time possess the capabilities required to respond optimally to military-technological developments is declining.

Project description	Project volume	Estimated expenditure in € million						Phasing
		t/m 2014	2015	2016	2017	2018	2019	
Royal Navy								
Evolved Sea Sparrow Missile Block II: participation in international development process	37,2	35,5	1,7					2015
Maintenance of M-frigates	58,7	53,2	2,8	2,7				2016
Maintenance of Walrus class submarines	96,0	42,2	15,6	9,1	7,6	8,1	6,8	2020
Maintenance of Goalkeeper	34,5	20,5	7,7	6,2				2016
Low Frequency Active Sonar (LFAS)	27,4	22,6	2,5	2,3				2016
Air Defence and Command Frigates	1560,3	1553,8	6,5					2015
Maritime Ballistic Missile Defence (MBMD)	124,6	60,4	22,9	13,0	15,6	6,4	6,3	2021
Patrol ships	529,9	522,7	7,2					2015
Upgrade MK 48 torpedo	71,8	24,0	15,8	16,1	15,9			2017
Acquisition of Joint Logistics Support Ship (JSS)	409,3	379,9	29,4					2015
Royal Netherlands Army								
Army Ground Based Air Defence System (AGBADS)	126,3	126,3						2015
Battlefield Management System (BMS) and Datacomunication for Mobile Deployment (DCMO)	BMS 62,8 DCMO 43,0	62,8 39,2	2,0	1,8				2016
PATRIOT replacement by COMPATRIOT	30,8	17,0	13,8					2017
Large Armoured Fighting Vehicle (GPW, Boxer), production	Productie: 794,4	470,7	132,3	126,9	53,4	11,1		2018
Infantry Combat Vehicle (IGV), production and training	1118,1	1116,6	1,5					2015
Replacement of utility and main battle tanks	90,5	85,	5,4					2015
Royal Netherlands Air Force								
AH-64D Block II upgrade	120,0	41,5	32,5	41,0	5,0			2013-2017
AH-64D weapons upgrade	25,9	2,7	9,6	12,0	1,6			2013-2017
AH-64D self-protection (ASE)	76,1	12,4	25,3	34,4	3,9			2013-2017
Chinook expansion and reinforcement (four + two)	356,2	351,5	4,7					2007-2015
F-16 infrared guided air-to-air missile	31,9	2,3	15,3	14,3				2013-2016
F-16 M5 modification	38,8	36,1	2,7					2015
F-16 mode 5 IFF	39,7	26,4	4,0	9,3				2016
F-16 air-to-ground weapons upgrade, phase 1	59,1	52,7	6,4					2004-2015
F-16 air-to-ground weapons upgrade, phase 2	75,3	25,3						2012-2027
F-16 self-protection (ASE)	82,0	32,5	25,5	24,0				2009-2016
Lengthening F-16 Flight Time – Flight Safety & Airwothiness	37,0	2,4	6,7	7,6	7,3	6,5	3,5	2014-2021
All military branches								
Counter Improvised Explosive Devices (C-IED) block 1 & 2	29,6	27,4	2,2					2015
Counter Improvised Explosive Devices (C-IED) block 3 (structural embedding)	53,6	6,6	13,7	11,1	5,1	6,6	8,1	2022
Military Satellite Communication, long-term for all branches (MILSATCOM)	132,1	121	9,0	1,8	0,3			2017
Military Satellite Capacity (MILSATCAP)	31,4	16,1	7,5	5,3	1,4	1,1		2018
Modernization of navigation systems	39,0	20,9	3,4	2,5	6,0	2,2	4,0	2019
NH-90	1197,4	964,8	70,9	72,1	64,1	25,5		2018
Expansion of Chemical, Biological-Radiological and Nuclear (CBRN-) capacity in the context of the Intensification Civil-Military Cooperation (ICMS), materiel	60,3	21,3	20,8	18,2				2016

Fig. 6.5 Planning overview for the materiel of the armed forces. (Source: Ministry of Defence)

6.4 Future-Proofing the Armed Forces: The Dilemmas

According to the WRR,[13] the armed forces need to move from "doing a little of everything" to "focus".[14] That focus is tightening, gradually. By formulating policy emphases in the International Security Strategy, such as the orientation towards the countries around Europe, more intensive European cooperation, the integrated approach and prevention, a start has been made in the process of setting priorities. However, if prioritisation is also taken to mean the concentration of resources – policymaking, financial, diplomatic, military and otherwise – on the selected policy priorities, there is still a long way to go. Although since 2014 the government has, in light of the worsening security situation and with broad support from the House of Representatives, made additional funds available, rising to 870 million euro in 2021, the measures that have been taken primarily address the most serious problems facing the armed forces. In light of the greater instability around Europe and global developments, politicians are faced with important questions in making the armed forces more future-proof.

Following on from the *Marinestudie* [Navy Study] in 2005, in the last decade the navy has focused more on providing support for land-based operations and, on the assumption that the threat at sea has diminished, has opted for Ocean-going Patrol Vessels (OPV) at the expense of more heavily armed multipurpose frigates. Furthermore, little has come of the planned strengthening of the land-oriented capabilities. In the context of *In the interest of the Netherlands*, for example, one of the tasks of the Joint Support Ship, the function of providing a helicopter platform for the Marine Corps, was abandoned.

A more urgent question, however, is what impact the altered security environment is having on the size and composition of the naval fleet. The ability to control (access to) parts of the seas has become topical again in light of Russia's greater assertiveness in European waters. At global level, the maritime arms race in Asia stands out. Mine clearance, anti-submarine operations and maritime surveillance closer to home have also become far more important. The purchase of additional maritime capacity seems logical given the increasing importance – also for the Netherlands – of flow security. Combating piracy and protecting the landing points of transatlantic cables and, last but not least, protecting the Caribbean parts of the Kingdom are further important arguments for expanding the navy's capacity. However, it is extremely doubtful whether even the planned replacement of capabilities with effect from the coming decade, including mine hunters, multi-purpose frigates and submarines, will be possible within the current budgets.

The army has become more mobile. This is appropriate in the context of the requirements of hybrid warfare and smaller crisis management operations, but presumes further steps, such as the further strengthening of the Commando Corps, to meet the growing demand for special forces. At the same time, the return of the threat of conventional war in Eastern Europe requires further strengthening of conventional capabilities, such as the return of a modest number of tanks, albeit in a German tank battalion, which will in turn contribute to the army's further integration into the *Bundeswehr*. Quite apart from that, the army in particular faces the task of restoring the balance between combat and support units.

The air force, finally, will at present have fewer F-35 fighter planes than originally planned because of the financial constraints. The changed security environment and the severely reduced sustainability are serious arguments for still considering the purchase of additional fighter planes. There are also good reasons for reconsidering the postponement of the purchase of the MALE UAV (unmanned aerial vehicles). This is a capability that is much in demand and an established military shortfall in Europe. It also seems logical to enhance the possibilities for innovation, in particular with respect to control of the information domain and the militarisation of space.

Do we opt for more F-35 fighter planes? With the number currently planned, only a small number of planes (four) will be available for use in operations. Do we buy extra frigates and submarines? That seems reasonable given the growing concerns about flow security and the navy's greater orientation towards coastal waters and supporting land-based operations. Or do we strengthen the army? After all, there is a need for more firepower, greater mobility and wider capabilities in conventional, irregular and hybrid scenarios. Additional requirements for the armed forces as a whole have also been identified in the domain of intelligence gathering and analysis, cyber and transport (helicopters). NATO has also identified various requirements, some of them the same (NATO Defence Planning Capability Review 2015/16: The Netherlands). The Netherlands also has responsibilities in that respect. Every choice the Netherlands makes or wishes to make will have to be reviewed in consultation with its partners in the EU and NATO. But that does not mean that the Netherlands cannot formulate its own strategy for the national armed forces, which it can very easily do in consultation with the allies and in alignment with the security strategies of the EU and NATO. Major advances can be made in that respect, particularly at European level.

6.5 Investing in Future-Proofing the Armed Forces in the EU

In the official response to the WRR-report (2017) on behalf of the Council of Ministers, the Minister of Defence states: "The Netherlands and Europe are confronted with a complex, diverse and uncertain threat assessment. Our interests and values are therefore at stake".[15] According to the Minister of Defence, "collaboration with the aim of strengthening each other and increasing employability through far-reaching interoperability is part of the collaboration that is being sought".[16] Therefore, "in the European context, cooperation must become the norm and no longer the exception".[17]

Progress in expanding and deepening the cooperation within the EU is important for tightening the focus of the armed forces. There is no question of forming a supranational European army or of the complete integration of the European security and defence policy in NATO.[18] But Europe's capacity to act can be enhanced with pragmatic cooperation (*Rapportage internationale militaire samenwerking*

[Report on international military cooperation][19]). Such cooperation could relate to the procurement of weapons systems (Ministry of Finance/IBO 2015) or deployment in specific operations. Other general forms of cooperation are possible, ranging from one-off (such as joint exercises) to permanent integration (for example, the German-Netherlands Corps headquarters). Figure 6.6 presents a brief history of European defence cooperation.

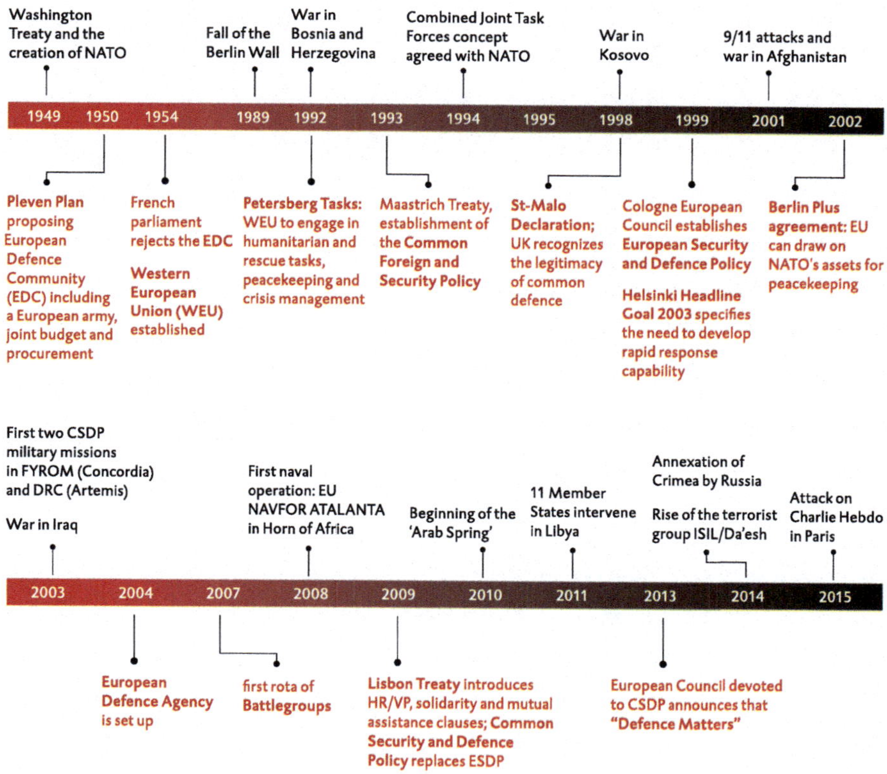

Fig. 6.6 The long road to European defence. (Source: EPSC Strategic Notes, June 2015)

Potential partners are countries whose political/geopolitical strategies are close to those of the Netherlands and that have a similar political culture to the Netherlands (strategic partners), countries with which the Netherlands cooperates within NATO and the EU (allied partners) and other countries that do not fall into either of the first two categories (ad-hoc partners). Research into examples of cooperation has identified a number of success and failure factors.[20] It has already been mentioned that Europe is not pulling its weight with its contributions to defence. Figure 6.7 shows what that means for each category of expenditure.

The decline in expenditure on investment and research is particularly worrying for the future. Figures 6.8 and 6.9 provide a breakdown.

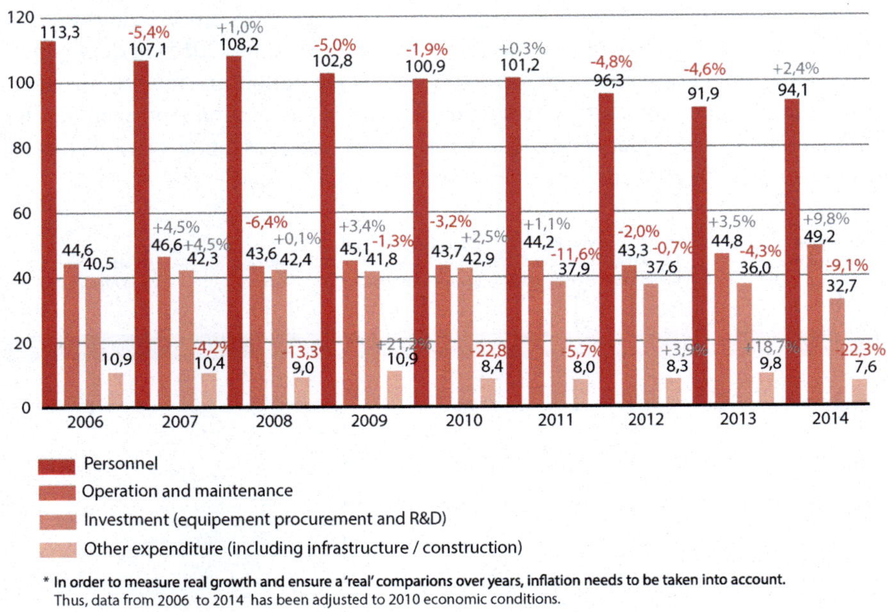

Fig. 6.7 Defence spending in the EU, in 2010 euros. (Source: European Defence Agency Data)

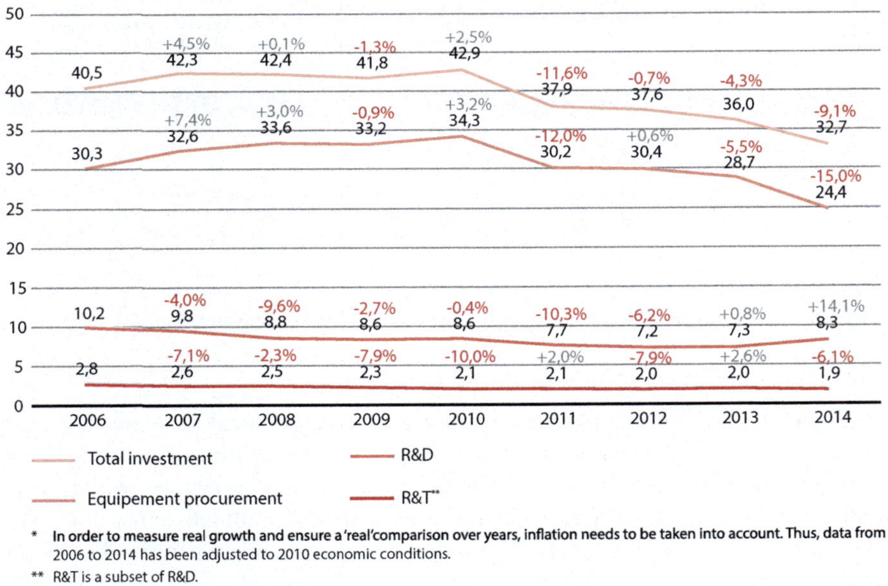

Fig. 6.8 Expenditure on investment in the EU. (Source: European Defence Agency Data)

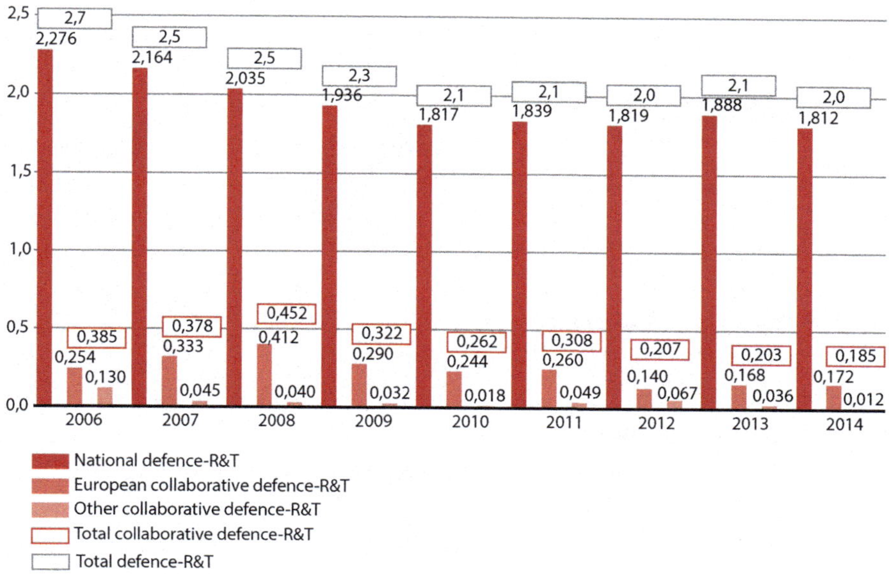

Fig. 6.9 Expenditure on R&T in the EU. (Source: European Defence Agency Data)

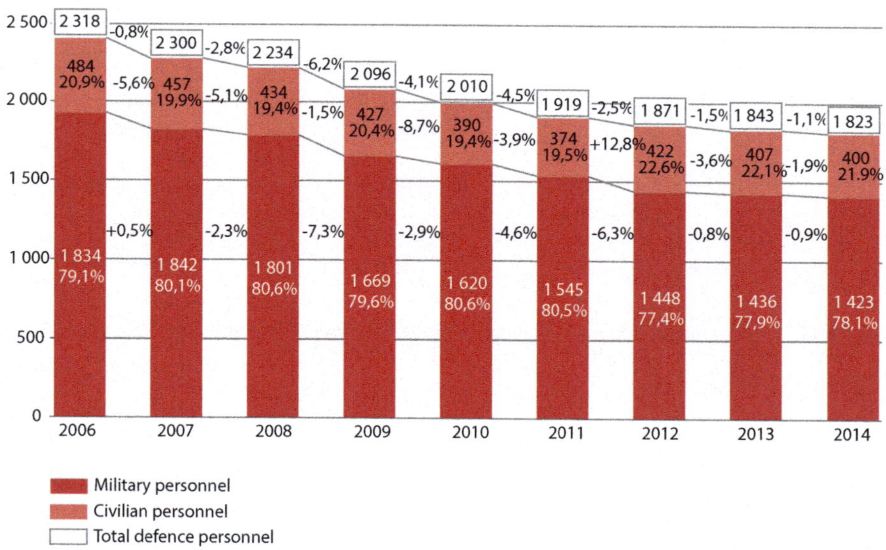

Fig. 6.10 Military and civilian personnel in the EU. (Source: European Defence Agency Data)

Knowledge and investment in new weapons systems cannot be seen separately from the personnel requirements (see Fig. 6.10). Demographic developments, the demand for highly skilled personnel and the need to retain personnel are major challenges, which have to be considered in relation to the investments in research.

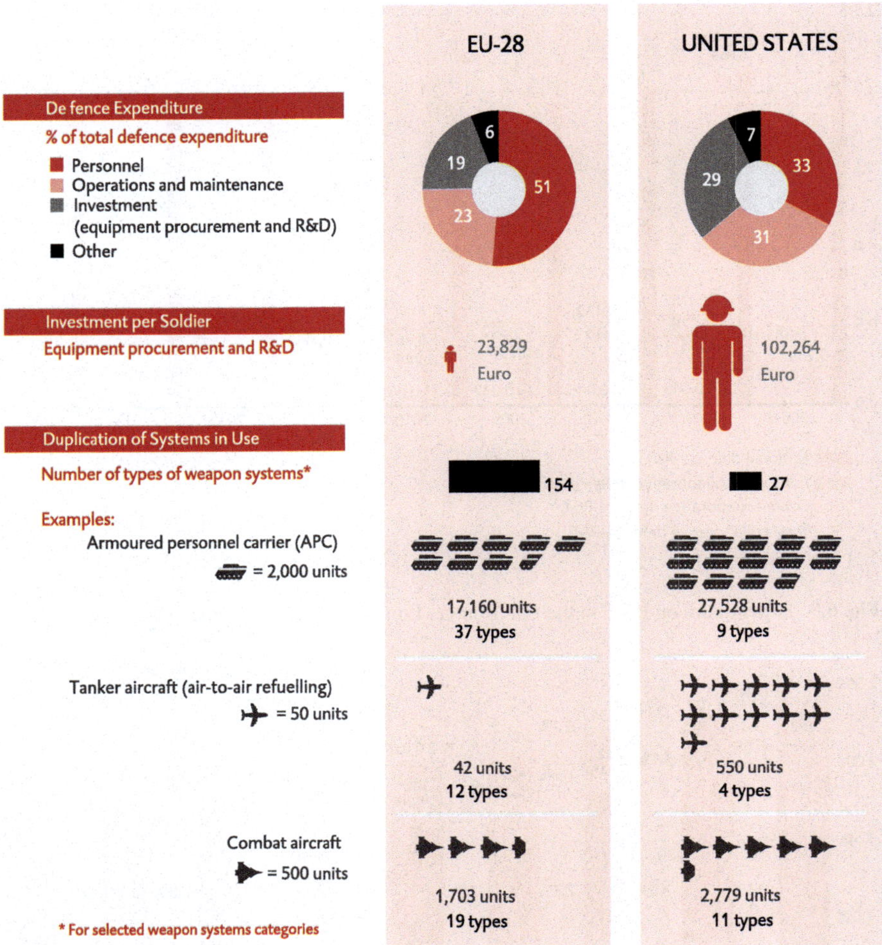

Fig. 6.11 Lack of integration in EU defence in figures. (Source: EPHS Strategic Notes, June 2015)

As Fig. 6.11 shows, the member states of the EU could collaborate more. It has been calculated that the lack of integration between national military structures and the absence of an integrated defence market costs at least 26 billion euros a year.[21] The variety of weapons systems causes fragmentation and loss of efficiency due to insufficient economies of scale and weakens negotiating power vis-á-vis manufacturers (see Fig. 6.12). In times of crisis, the absence of uniformity will lead more quickly to shortages of replacement materiel, which is an unnecessary source of vulnerability for European defence.

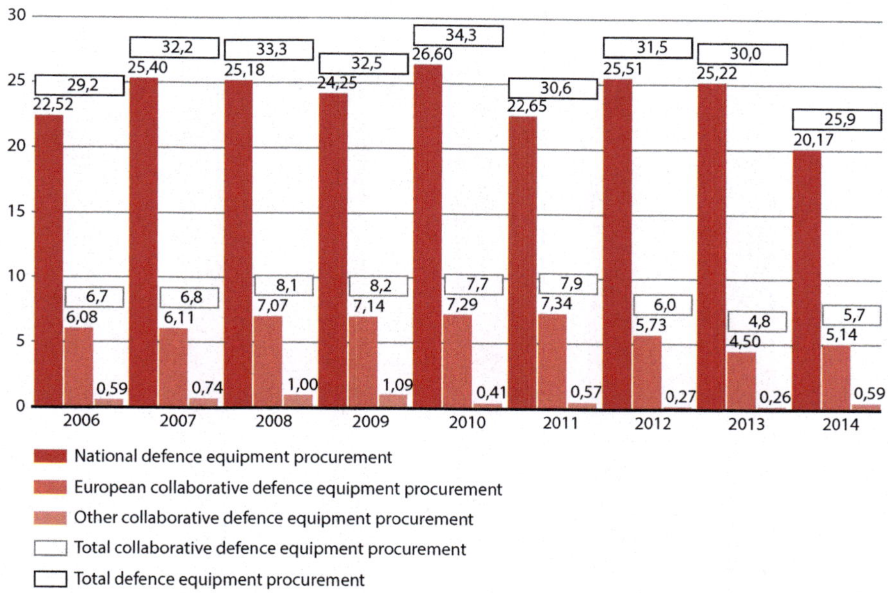

Fig. 6.12 National and multilateral expenditure in the EU. (Source: European Defence Agency Data)

The actual multilateral cooperation extends to just a small percentage of the total budget (see Figs. 6.12 and 6.13) and the trend is in fact slightly downwards. The Interdepartmental Policy Study entitled *Meer Bang for the Buck [More Bang for the Buck]*[22] does argue, however, that bilateral cooperation in the procurement and development of new materiel is often easier than multilateral cooperation.

The larger the number of countries and national defence industries involved, the more the requirements will vary over time and/or substantive differences will arise. And the less mutual trust there is, the greater the chance that the cooperation will suffer at the expense of value for money. This should be an important guideline for future projects.

The Dutch armed forces have a good reputation when it comes to military cooperation within the alliances. As regards the army, the cooperation with Germany has considerable political and military significance. The same applies for the collaboration between the navies of the Benelux countries in BENESAM, in which the Netherlands is the senior partner. Another example of successful cooperation is the joint protection of the Benelux air space. The possibilities for military cooperation are far from exhausted if use is made of joint operational requirements documents, an Interdepartmental Policy Study (IBO) concluded. It further argued that closer military cooperation will facilitate the retention of economies of scale, which is essential when costs are climbing and volumes are falling (cf. *Meer Bang for the Buck*, an IBO on ways of enhancing effectiveness through international cooperation and integrated contracts for weapons systems for the Ministry of Defence, Ministry of Finance, The Hague, April 2015).

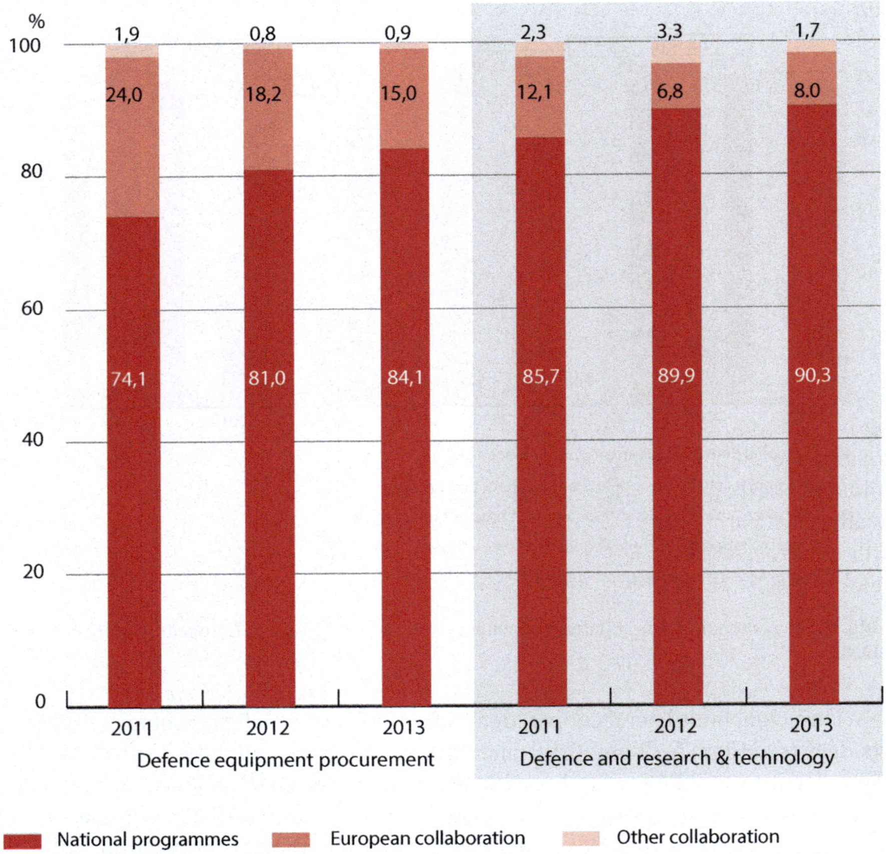

Fig. 6.13 European cooperation, still the exception rather than the rule. (Source: EPSC strategic notes, June 2015)

In that context, it is advisable to make joint procurement and maintenance a condition when replacing capabilities and to abandon the policy of retaining basic and niche capabilities. However, there will still be restrictions. The need to perform national tasks independently and the political desirability of being able to act without a partner or with a different partner if possible are natural constraints in this respect. That in fact applies not only for the Netherlands, but for every European country with armed forces of some size.[23] The risk of surrendering sovereignty is not necessarily an obstacle to this; in 2012 the government embraced the advice of the AIV[24] to define sovereignty as the capacity to act and in that light to regard military cooperation as strengthening rather than impairing the Netherlands' capacity to act. However, the Netherlands is by no means in the vanguard when it comes to helping to reduce Europe's identified military shortcomings. In 2008, the Netherlands withdrew from NATO's multinational Air Ground Surveillance (AGS) project. The MALE UAV project (unmanned aerial vehicles with a substantial range) has been postponed. These and other European shortfalls involve essential instruments/systems that are crucial for strengthening Europe's capacity to act, for retaining US involvement in European security and, equally, the capacity of the Dutch armed forces to act. The Netherlands should make amends and, preferably in collaboration with partners and allies, prepare plans and earmark money for the procurement and operation of these critical capabilities.

6.6 The Path to a Tighter Focus and Additional Investment

Maintaining a high-class expeditionary armed forces and a willingness to continue investing in it are important requirements for sustaining the ability to continue making relevant military contributions to NATO, the EU and the UN and in ad-hoc coalitions. However, the return on these investments, in other words the relevance of future military contributions, will be determined by the extent to which the Netherlands and its allies and partners are able to strengthen NATO and the EU's capacity to act, since, with the exception of national tasks, there is no question of acting alone. Member states can only address the growing instability around Europe together.

Despite NATO and EU initiatives to address the identified shortfalls in a coordinated fashion, the capacity to act jointly, and without the US if necessary, is still under-developed. For example, there is a lack of (adequate) strategic instruments, crucial support capabilities such as reconnaissance equipment, tactical and strategic transport equipment, capabilities for aerial refuelling, the right types and quantities of precision ammunition and sufficient firepower for actions across the entire spectrum of the use of force in general.

In the last few decades the Dutch armed forces have evolved into a professional expeditionary force, with highly skilled personnel and modern capabilities. Their repeated deployment for stabilisation and peacekeeping operations with NATO, the EU and the UN and in coalitions underlines the development of the armed forces into an important instrument of foreign and security policy. However, this process of modernisation has been financed from a shrinking budget, while operating and investment costs and the costs of wear and tear have risen as a result of demanding missions. This approach has caused deep-seated problems. Deployability is under pressure due to shortages of ammunition, spare parts and technical staff and possibilities to provide adequate training for troops. The choice made in the austerity programmes to favour combat strength over support also imposes structural constraints on deployability. Added to the greatly diminished redundancy within the armed forces – the numbers of units and capabilities have declined greatly as a result of the reduction in the size of the armed forces – it is questionable whether the armed forces could respond adequately in the event of additional and simultaneous calls being made on them – in this country or elsewhere. The armed forces must therefore be regarded as insufficiently future-proof.

These problems have taken on a different aspect with the deterioration of the security environment around Europe, the agreements in Wales (2014) and Warsaw

(2016) to strengthen the alliance's defences and increase defence spending and the growing vulnerability of the Netherlands and its inhabitants.

Making the armed forces more future-proof will take more than a phased strengthening. The capacity to respond to the changes in the security environment and military-technological developments is limited at the moment. The current defence planning is dominated by the replacement of existing weapons systems. Too few people and resources are assigned to knowledge, innovation and modernisation of the armed forces. The same applies for reducing the identified European military shortfalls. Up to now, the Netherlands' contribution to addressing these issues has been very limited, despite the crucial importance of these capabilities for Europe's capacity to act. This is not an exclusively financial issue, as *Houvast in een onzekere wereld. Lijnen van ontwikkeling in het meerjarig perspectief voor een duurzaam gerede en snel inzetbare krijgsmacht* (2017) has shown. The defence planning system should adopt a more flexible and focused division between the army, the air force and the navy to facilitate integrated, future-oriented deliberations.

The Netherlands is therefore faced with important questions regarding its security. How can the growing instability around Europe be reduced? What role should NATO and the EU play in that? What specific contributions can the Netherlands make? Which elements will be given priority? And what additional resources can and will the Netherlands free up in light of the deteriorating security environment?

Additional steps are needed to tighten the focus in the deployment of resources. The political debate should create transparency about the detailed choices that the armed forces have to make to tighten the focus. The aim of this book is to stimulate the debate among politicians, policymakers and strategists on this issue. The following closing chapter presents the final conclusions and recommendations in which, to supplement the foregoing analysis, five areas of concern are presented to help determine the necessary focus. These areas of concern create the conditions under which the conclusions and recommendations presented in the concluding chapter should be read:

(1) a more strategic embedding of defence policy;
(2) the establishment of greater anticipatory capacity;
(3) the strengthening of the defence planning process on the basis of a strategic vision;
(4) guaranteeing a stable multi-year perspective;
(5) creating more room for knowledge and innovation.

6.6.1 A More Strategic Embedding of the Armed Forces

The choice of capabilities of the armed forces in the context of the cooperation in the NATO and EU alliances requires strategic embedding. The Netherlands always works with partners, but also needs its own strategy. The International Security Strategy and the National Security Strategy currently lack the necessary focus. Strategy documents like the International Security Strategy and the National Security Strategy and coalition agreements invariably contain relevant passages, but texts that express values and interests in general terms do not automatically compel the making of choices. Vague formulations are not very useful. The AIV, the WRR and the Clingendael Institute have therefore all recently produced more precise definitions of interests and values and incorporated them in an assessment framework.[25] Conclusions can also be drawn from the foregoing analysis of strategic trends, threats and risks.

From the perspective of the armed forces, the preferred option is an integrated security strategy that also compels the other 'client' ministries to coordinate their priorities and direction. After all, the needs of the ministries of Foreign Affairs and Justice and Security directly affect the targets for the deployability of the armed forces. It also fits in with the steadily expanding interconnectedness of internal and external security.

6.6.2 The Establishment of Greater Anticipatory Capacity

The security strategy must penetrate deep into defence policy to make a real difference to it. The final report of the interdepartmental *Verkenningen: houvast voor de krijgsmacht van de toekomst* [Future Policy Survey. A New Foundation for the Netherlands Armed Forces] (2010) recognised anticipation, defined as preparing for foreseen and unforeseen developments and incidents that may affect the interests of the Kingdom of the Netherlands or the international rule of law, as one of the Dutch government's strategic security functions. Like the recent French security strategies, the *Livre Blancs* in 2008 and 2013, it emphasised the need for an adequate strategic intelligence position and constant monitoring. The same applies for the importance of knowledge ("maintaining a sufficiently broad and relevant knowledge base"). Dutch priorities relate in particular to the need to draw up scenario analyses and foresight studies in the context of a future-oriented policy and capacity development with which sufficient flexibility, adaptability and resilience in the defence organisation can be guaranteed. In that context, the Future Policy Survey advocated (referring to it as a "policy consideration") permanently embedding the foresight method in the Ministry of Defence's regular policy-making process. That proposal was not followed at the time. In contrast to countries such as Germany, France and the United Kingdom, the Ministry of Defence still does not possess such a capability to support the internal policy and capacity development process. The perseverance with strategic foresight studies in the form of the Strategic Monitor is valuable, but is not in itself enough. Updating the regular scenarios and related studies produced by the HCSS and the Clingendael Institute has only limited value as long as the Ministry of Defence lacks the capacity to analyse them and they have no measurable effect on policy and capacity development. Consequently, the Ministry of Defence is failing to take advantage of opportunities to accelerate the pace at which it adapts (or upgrades) operational capabilities in response to the changing environment. The process of strategic foresight should be permanently embedded in the defence organisation in order to transform the Ministry of Defence into a strategically agile organisation.[26]

6.6.3 Strengthening of the Defence Planning Process

The Netherlands can use the strategy formulation process to enhance defence planning. The international coordination, particularly with a view to international cooperation (NATO, EU), could also be improved. In the Defence Materiel Process, the House of Representatives concentrates on the results of the planning process when the choice of a particular capability has been made. The operating costs of the capabilities to be procured receive a lot of attention in that process. After all the cost overruns and shortfalls, the Netherlands Court of Audit also concentrates on that aspect. A lot has been written in the Netherlands about the capabilities the Netherlands should possess, but almost nothing has been published about defence planning. Capabilities are still assessed individually for their usefulness and versatility; there is no integrated review of their relative effectiveness and efficiency in achieving the desired military effects. Added to the requirement that the capabilities must fit in with the financial planning and must not cause displacement effects (must not be at the expense of other capabilities), it is impossible to set priorities in favour of an Operational Command, a joint capability or an established European military *shortfall*.

First of all, there is a need for long-term defence planning that provides insight into future risks, promotes thinking in terms of scenarios and produces strength-weakness analyses in a changing security environment.[27] That is also essential for setting priorities.

The current system is inevitably dominated by the replacement of main weapons systems. Long-term defence planning could be further improved by creating anticipatory capacity within the Ministry of Defence.

Secondly, there is a need for close coordination with NATO and the EU, the most important clients of Dutch military capabilities. At present, the NATO Defence Planning Process and the EU's *Capability Development Plan* do not play a significant role in the choice of capabilities. However, by ignoring the identified European shortfalls, the Netherlands could be hurting itself, since they involve so-called support that individual countries cannot afford or maintain themselves. Thirdly, there is a need for strategic guidance of the principal clients of defence products, the ministries of Foreign Affairs and Defence. The choice of capabilities is primarily a political issue. Other ministries have traditionally been excluded from this process in the Netherlands, in contrast to countries like the United Kingdom, France, Finland, Denmark and Sweden, where parliaments are closely involved in the decision-making.[28] NATO's best-practice approach offers various starting points for a reappraisal of the current defence planning system and for refining the point of departure of the Dutch defence planning system, planning on the basis of the possibilities.

6.6.4 Guaranteeing a Stable Multi-year Perspective

The strategically enhanced defence planning process for the Dutch armed forces depends on a stable multi-year perspective, which also make provision for sufficient resources to sustain the armed forces in the longer term. However, the long-term orientation is not helped by the way in which the Dutch defence budget is drafted up.

The defence budget is the result of an annual cycle that ends with the passing of the act containing the budget of the Ministry of Defence by the States-General. The budget can also be revised in the course of the year with supplementary budgetary legislation. This flexible budgetary mechanism allowed defence spending – as a percentage of GDP – to be halved in the decades after the end of the Cold War. In 2015, total defence spending came to 1.16% of GDP, which is well below the target of 2% agreed at the NATO summits in Wales (2014) and Warsaw (2016). The investment ratio in particular has suffered under the successive rounds of cutbacks. The actual investment ratio in 2015 was 15%, which is well below the NATO standard of 20% of total defence spending (see the letter from the Minister of Defence to the House of Representatives on the development of the investment ratio of the Ministry of Defence, 14 March 2016).

Although a number of planned cutbacks have been reversed in the last few years and some 'windfalls' have been spent on defence, these recent measures confirm the impression of a lack of consistency in budgetary policy. There are therefore growing calls to make the budgetary system less susceptible to political opportunism and fluctuations in the short term (see the motion by Eijsink, Parliamentary Documents II 2015–2016 34,300 X, no. 49, 12 November 2015). Annual fluctuations in the order of 20% of the total defence budget are not uncommon. That makes it difficult to plan and creates uncertainty within the organisation.

In the official response to the WRR-report (2017), the Minister of Defence states that the percentage of GDP that is spent on Defence will increase from 1.17% in 2017 to 1.26% in 2021. The Council of Ministers is aware that spending on Defence is far from 2% of GDP and that the European average has not yet been achieved. The of Council of Ministers aims to put in place "longer-term measures" ["*langere lijnen naar de toekomst*"] that are required for stable financing and reinforcement of the armed forces during this cabinet period. The Council aims to do so in the light of the NATO agreement of 2014 to move defence spending towards the NATO standard of 2% GDP in 10 years. It includes step-by-step growth in the context of these longer-term measures in order to achieve the capacity objectives of NATO in the review of the Defence Memorandum. This is planned for 2020. A possible extra follow-up step during this term of office will be examined in the light of the development of the security situation, the government-wide priorities and within the agreed budgetary frameworks (Parliamentary Documents II 2017–2018, 33 763, No. 141, 28 March 2018: 5).

Agreements on defence spending for periods of 5 or 6 years – separate from the sitting government's term of office – could create greater continuity and certainty. That is also important in view of the long-term investments that have to be made to ensure the defence forces are properly equipped in terms of personnel and materiel. But the discussion does not have to be confined to the question of whether or not a multi-year budget should coincide with a government's term of office (in other words, either 4 years or 5 or 6 years). A pragmatic approach is also an option, for example by working towards meeting the 2% target in 2024, 10 years after the agreements in Wales (2014), from the start of the new government's term of office.

Some countries already have experience with this type of multi-year agreement that fixes defence spending for a number of years. These agreements vary in terms of the period adopted, the degree of flexibility and the nature of parliamentary involvement. The Netherlands could in particular learn from the experiences in Denmark and Sweden (Drent en Meijnders, Multi-year Defence Agreements. A Model for Modern Defence?, Clingendael 2015; Bakker and Drent, *Meerjarige Defensie Akkoorden in Nederland* [Multi-year Defence Agreements in the Netherlands], Clingendael 2016). Creating an investment fund covering a period of 15–20 years – similar to the Delta fund or the Multi-year programme for Infrastructure, Space and Transport – could also help to create greater financial stability.

Successive cutbacks and under-investment have led to a reduction in the deployability and sustainability of the armed forces. The available resources are not even sufficient to meet the now greatly reduced ambitions (see Netherlands Court of Audit, *Resultaten verantwoordingsonderzoek 2015 Ministerie van Defensie* [Results of the audit of the Ministry of Defence 2015], May 2016; see also Schramade 2016). That is worrying in a drastically altered national and international security environment. Defence is, after all, the collective good *par excellence*. That special character justifies a certain depoliticisation of the defence budget, although the decision to do that is actually a form of self-policing and can only be a political decision. In short, multi-year defence agreements are essential for future-proofing the Dutch armed forces. They would also demonstrate to the international community the Netherlands' commitment to a course in which its own choices are naturally dictated in part by shortcomings in the alliance. Such agreements could therefore also contribute to the defence cooperation within NATO and the EU.

6.6.5 Creating More Room for Knowledge and Innovation

The Ministry of Defence will have to steadily accelerate the pace of innovation to keep pace with the dynamic in security challenges and to be able to effectively respond to technological and social developments (see Ministry of Defence, *Strategische Kennis-en Innovatieagenda 2016–2020* [Strategic Knowledge and Innovation Agenda 2016–2020], October 2016). The changing security environment and, more generally, the increasingly rapid pace of technological and social changes, call for innovations in the operational, doctrinaire, personnel and technological domains. The armed forces, and in particular the navy, have a good reputation for incremental innovation. However, replacing and upgrading platforms do not provide sufficient assurance of being able to keep pace with technological and operational developments in the long term. The possibilities of joining in new and possibly groundbreaking developments, such as nano technology and 3D printers, are very limited at the moment and therefore need to be expanded. The same applies for efforts to develop and test central concepts, methods and resources more quickly and more cheaply within the armed forces.[29] However, successful innovation still depends on a robust knowledge base and structure within the Ministry of Defence. The litmus test for successful innovation remains the choice of new weapons systems, technologies and concepts in accordance with the chosen strategy.[30] The explanation for this is that armed forces rely heavily on available and proven capabilities when it comes to the deployment of troops.[31] Most armed forces, including the Dutch, are no exception to that rule. They concentrate on incremental innovation. There is therefore less chance of more radical or even disruptive innovation. However, in the longer term the availability of new capabilities could be decisive for gaining the upper hand in conflicts.

The changes that defence policy has to address and the necessary strengthening of the armed forces must reflect the changes in the nature of warfare and make better use of the possibilities for European cooperation, bearing in mind the primacy of the framework of the NATO alliance. A sharper focus and increased investment are needed, however. The five steps set out above indicate the policy path to achieving that.

Endnotes

1 Wijk, R. de (2010).
2 Korteweg, A.R. (2011).
3 Zandee, D. (2016).
4 Beeres, R. & Bakker, E.-J. de (2010).
5 Lindley-French, J. & Tjepkema, A. (2010).
6 Hoffenaar, J. (2009: 14–15).
7 Beek, ter (1996).
8 Parliamentary Documents II, 33763, no. 57.
9 Parliamentary Documents II, 33763, no. 74.
10 Raynova, D. & Kearns, I. (2015).
11 Parliamentary Documents II, 34000, no. 23.
12 Parliamentary Documents II, 27925-506, 24 September 2014.
13 WRR. (2010a, b).
14 WRR. (2010a, b: 55).
15 Parliamentary Documents II 2017–2018, 33 763, No. 141, 28 March 2018: 6.
16 Parliamentary Documents II 2017–2018, 33 763, No. 141, 28 March 2018: 4.
17 Parliamentary Documents II 2017–2018, 33 763, No. 141, 28 March 2018: 4.
18 Andersson, J. et al. (2016).
19 Parliamentary Documents II 2015–2016, 33 279 no. 16.
20 Zandee, D., Drent, M. & Hendriks, R. (2016).
21 Ballester, B. (2013).
22 Ministry of Finance (April 2015).

23 Vries, S. de (2014).

24 AIV. (2012).

25 In 2010, the WRR concluded that Dutch foreign policy consists of "a broad range of aspirations, viewpoints and activities" and that "in the middle of this range there is little to connect the various elements". The report compared Dutch foreign policy with a "doughnut". There was a risk that "policymakers and policy followers will not be able to see the wood for the trees, with all the negative consequences this will for the results, for legitimacy, for areas of expertise and for internal and external authorities". That called for a transparent assessment framework and the setting of priorities. In 2013, in a study into the armed forces of the future, experts from the Clingendael Institute distinguished four interests: (1) influence in the international community; (2) preservation of prosperity and economic development; (3) enhancing security and stability; and (4) promoting human rights and humanity. They suggested an appropriate armed forces for each of these interests: (1) an air-based intervention force, (2) a maritime force, (3) a robust stabilisation force and (4) a supporting peace force (Colijn et al. 2013). The experts submitted all four options to the politicians, with interest 3 (promoting security and stability) and option 3 (a robust stabilisation force) as their own preferred choices. The policy document In the interest of the Netherlands shows that the second Rutte government chose neither of these options, but continued to follow the course taken by previous governments, that of versatile deployable armed forces.

26 Bekkers, F. & Spiegeleire, S. de (2010).

27 Stojkovic, D. & Dahl, B.R. (2007); Larsen, K. & Håkenstad, M. (2012).

28 The Ministerial Committee on the replacement of the F-16 was specially established to make a final assessment of the plans for the replacement of the F-16 and the calculations of the affordability of the 29 main weapons systems of the armed forces.

29 Meessen, W. & Wiel, W. van de (2014).

30 Wagen, C.M. (2012).

31 Murray, W. (2001).

Chapter 7
Interconnected Security: Conclusions and Recommendations

Ernst Hirsch Ballin, Huub Dijstelbloem, and Peter de Goede

7.1 Security, Defence, Armed Forces: The Consequences

Strategic analysis, security policy, defence policy, the armed forces: that was the sequence followed in this book. The same sequence should be followed for the purpose of policy analysis. On the basis of a strategic analysis of the altered security environment, this book has considered what defence policy is required now and in the near future and where the armed forces need to be strengthened. The emphasis was on where the focus needs to be tightened and how additional investment should be used. The precise choices and priorities that should be formulated in that process must be determined in the political debate. It is of the utmost importance to conduct that debate. Security policy and defence policy depend on it. This final chapter makes a number of recommendations on how to make the necessary improvements and to create a more strategic foundation for the thinking of politicians and policy-makers on security and defence. The book concludes with a discussion of the following points:

- the consequences of 'security in an interconnected world', with an evaluation of the background to the current security and defence policies and suggestions for the course to be followed in improving policy (Sect. 7.2);
- the conclusions and recommendations (Sect. 7.3).

E. Hirsch Ballin (✉) · H. Dijstelbloem · P. de Goede
WRR, Scientific Council for Government Policy, The Hague, Zuid-Holland, The Netherlands
e-mail: ballin@tilburguniversity.edu; h.o.dijstelbloem@uva.nl; peter.goede@minbzk.nl

E. Hirsch Ballin et al. (eds.), *Security in an Interconnected World*,
Research for Policy, https://doi.org/10.1007/978-3-030-37606-2_7

7.2 Security in an Interconnected World

7.2.1 Freedom, Security and Sovereignty

Defence, the protection of the country, is a core task of the sovereign state. National sovereignty ultimately means that the organs of the state freely decide on the shape of the society and on the relations with other states for – and in a democratic state genuinely on behalf of – the inhabitants of that state. States are not only entitled, but have an obligation towards their population to defend that sovereignty against internal and external threats, alone or in an alliance with other states. Defence is therefore at the service of the collective security and political freedom of Dutch society against external threats. The Charter of the United Nation's recognises that states have the right to defend themselves individually or collectively against armed aggression.[1] However, every state is obliged to prevent such situations from arising. The international legal order contains rules for peaceful cooperation between states and the peaceful resolution of disputes. The international legal order provides the frameworks for cooperation between states through treaties and international or supranational organisations.

The sovereignty of states has formed the basis of the international legal order for centuries. That has not changed. What has changed is that it is no longer the sovereign that is seen as having the primary interest in the protection of freedom, but the society as a whole. Sovereignty now implies responsibility – internally for the security and the living conditions of citizens, and externally for peaceful relations with other states.[2] 'National security' is no longer seen as separate from, but as the prerequisite for the security of individuals and the society.

The realisation that preventing conflicts starts with the internal relationships only arose when repressive, contemptuous regimes unleashed the Second World War. The US President Franklin D. Roosevelt clearly formulated that insight in 1941, even before the US became directly involved in the war, in a speech in which he articulated the 'Four Freedoms': freedom of speech, freedom of worship, freedom from want and freedom from fear. Oppressors who violate those fundamental freedoms threaten their own people and international peace. For that reason, since the Second World War the UN and organisations like the Council of Europe have focused on fortifying human rights, living conditions and democratic freedoms. Many past and present international conflicts have arisen from internal conflicts and civil wars. Syria, as well as numerous acts of international terrorism, are dramatic examples of that. International security also starts with avoiding tensions, for example by creating a peaceful climate within national societies by preventing groups from being set against one another.

On the other hand, the freedom and security of states in relation to each other is one of the conditions for freedom and security *within* states: "All peoples have the right of self-determination. By virtue of that right they freely determine their political status and freely pursue their economic, social and cultural development." The two conventions in which the United Nations has laid down the protection of human rights both open with this contemporary definition of national sovereignty.[3] However much one tries to contain the methods of warfare, time and again the civilian population suffers, perhaps even more severely each time. After a conflict, it can take a generation or longer for the scars in the relations between peoples and ethnic groups to heal.[4]

7.2.2 The Facets of the Current Security Environment

Today, in 2017, it is perfectly clear how tightly the personal and collective freedom and security of the inhabitants of our state are entwined with international security. Crime, accidents and disasters create insecurity, but people regard terrorist violence as a far more serious threat to their security. Just as, in fact, the governments of totalitarian states threaten the liberty and security of their citizens. EU member states and NATO allies bordering on conflict zones face military threats, a risk that also cannot be ignored for the Caribbean parts of the Kingdom. In one way or another, every threat to the security of the countries of the Kingdom arises from situations in which the 'four freedoms' are being seriously impaired and sovereignty is regarded as a licence to act arbitrarily. It is therefore essential that security strategies as recommended in this book are permanently backed up and reappraised with thorough analyses of the environment.

The research described in this book confirms that security, seen from the perspective of the individual, of a state or of a society as a whole and from the perspective of inter-state relations, is interconnected. That applies first and foremost for the concept of human security, which has acquired a wide meaning in the thinking about development issues. It encompasses not only an individual's personal situation, but also the extent to which people can have faith in the government services and social conditions that are vital to them. The relevance of the 'four freedoms' is also apparent here. Human security is a subject that should ideally be studied in a multidisciplinary context: anthropological, geographic, psychological and legal, as well as the theory of international relations,[5] but also from the perspective of socio-economic conditions.

The term 'national security' is sometimes used in the sense of state security (the security of the most important institutions of the state; in other words, the domain of intelligence and security services), but more often the security of the nation state as a whole in the international context. External and internal threats to national security encompass not just disruption of the conditions under which people wish to live their lives, but also of the connections that people depend on. The unimpeded import and export of food, raw materials and other goods must be possible along connecting routes; the same applies for essential data traffic, which can be disrupted both physically (for example, by destroying maritime cables) and electronically (by online manipulation known as cyber warfare). All of these threats are covered by the term flow security.

To properly understand the issues facing contemporary security policy, it is useful to assess developments through the prismatic lens of these three facets – human security, national security and flow security. In that way, one avoids focusing solely on the current well-being of the country's own population in formulating strategies: at first glance that well-being seems to be best served by lower government spending on defence and development cooperation, while the basic conditions of that well-being might be threatened by international conflicts, which in fact call for international action, including becoming better equipped to protect the territorial integrity of the states belonging to the North Atlantic alliance and the EU. These facets also indicate that security policy is not just defence policy, but also affects foreign policy in a wider sense and, in this case, development cooperation in particular.

A prismatic consideration of international threats helps to prevent geopolitical security imaginaries from narrowing the analysis (as the reaction has been in recent years).[6] After all, what is happening on the other side of the alliances' external borders has an impact on the situation here, for example when refugees seek refuge in Western Europe. That is one of the consequences of the instability and fragility of states (fragility that can result in what are known as failing states). The consequences of that are not necessarily 'official' wars, but can also include actions by paramilitary groups, hackers and terrorists. All these manifestations of conflicts particularly affect the transport routes, trade flows and tourism that are vital to the Netherlands and its partners.

This prismatic analysis implies that the defence of the freedom and sovereignty of our state can no longer be broken down into internal and external. Nor can the defence of that freedom and sovereignty still be separated from those of other countries: alliances of free democratic states form the backbone of the international legal order. Dutch security strategies can no longer be drawn on a map, but can only be visualised in networks with geographic anchor points.

7.2.3 *Security Policy as a Facet of General Government Policy*

Dutch security policy can only be more effective if it responds to the international interconnectedness of our security environment. The Netherlands' security policy must therefore be part of North Atlantic and European security policy – in a manner appropriate to us and which reflects our own strengths and possibilities – or an overseas Caribbean security policy to be implemented in association with other states.

The perspectives outlined in this book and the recommendations it makes for improving the procedures and substance of security policy are based on insights into the origins of the current situation. They include the realisation that the most radical developments in the last quarter of a century were not anticipated. In 2002, a group of prominent researchers published a volume entitled *Internationale Politik im 21. Jahrhundert,*[7] containing chapters on all the relevant themes, such as global governance, humanitarian interventions, the EU, Russia and China. Only a small number of the expectations they expressed have actually come about.

The complex unpredictability of international relations is further magnified by a development that was not yet discussed in the review in the preceding chapters, namely climate change. The consequences of climate change could cause far greater migration streams, for example from the Nile Delta and the coastal area of Bangladesh,[8] than those from the semi-devastated Syria. There are also many other potential consequences, for example in terms of food security. This means that security policy can only be tailored to specific, known hazards to a limited extent. The recommendations made in this book are based on the assumption that Dutch security policy will have to be geared to European and NATO policies. That is not just a question of following them, but also contributing insights and complementary national policies, for example aimed at creating greater stability in other countries.

The three distinct facets of a prismatic analysis of developments identified in this book – human security, national security and flow security – can also be applied to the recommendations. What is clear is that the instability of the system of international relations arises from its multipolarity and the complexity of the relationships. Threats to flow security in the seas to the east of the Horn of Africa are fostered by the abject situation of the local population in terms of human security. It is, for example, a traditional task of the defence forces to protect Dutch vessels, while preventive policy would focus on improving the socio-economic situation in the region.

However, a security strategy calls for more than confirmation of what is already being done or could have been done. Capacity geared solely to the known risks will prove inadequate when unforeseen threats appear. For that reason alone, this book recommends rapidly raising the defence budget, but also the budgets for diplomacy and development cooperation, to bring them into line with the internationally accepted benchmarks. The UN's Sustainable Development Goals have a very wide scope; they could be placed under the proposed prismatic lens in order to determine how they could be relevant for reducing international tensions.[9]

A new security strategy will have to identify as clearly as possible which points in the international networks that are relevant to the Kingdom of the Netherlands belong to vital relationships. They could be situated in the Netherlands (for example, the landing points of transcontinental cables) or on transport routes, such as the seas around the Horn of Africa. They could also be countries of origin where internal tensions can be eased with peacekeeping missions and through social and economic development. For regions where the European Union and the Russian Federation and the European Union and the Arab countries are close neighbours, a neighbourhood policy that serves common interests should be developed.

The Kingdom of the Netherlands will have to coordinate the spending of defence funds with NATO allies and other EU member states. However, it is both sensible and legitimate to specify national preferences that match the country's own interests and capabilities. As a 'seafaring nation', the Netherlands could emphasise the strengthening of its navy, and in the process simultaneously provide an impulse for the maritime sector, including training for the merchant marine. The Netherlands' experience with missions in support of the police and judicial institutions and the demands of European border control are a valid reason to provide extra resources to strengthen the Royal Netherlands Marechaussee. The same applies for operations in the Caribbean region and for deployment in connection with flow security and the protection of vital infrastructure and hubs that are situated on the coast or at sea.

Together with six other NATO member states, the Netherlands has established the Civil-Military Cooperation Centre of Excellence in The Hague. The centre develops methods and provides training for civil-military cooperation in conflict situations, to protect cultural heritage and ecosystems, to strengthen the rule of law and to combat cyber crime and cyber terrorism. Strengthening this specialisation would reflect the observation that human security, national security and flow security are interconnected.

7.2.4 Monitoring, Understanding, Anticipating, Acting and Evaluating

This book recommends formulating an integrated security strategy that embraces internal and external security to the extent that they are intrinsically interconnected. It would not cover 'ordinary' crime, but should extend to combating international terrorism and cyber attacks by other states, since they represent attempts by organisations like Da'esh or by other states to disrupt Dutch society. A strategy will have to be constantly reappraised, which requires the capacity to monitor and understand developments and to tailor actions accordingly. A National Security Council at ministerial level should not only be provided with analyses by the intelligence and security services, but also have the support of academic research into international relations and political systems; the scientific policy needs to be strengthened in these respects.

7.2.5 Security and the Future of the International Legal Order

The question that needs to be addressed is how this reappraisal of the policy on international security relates to the Kingdom of the Netherlands' constitutional duty to promote the development of the international legal order. The international legal order does not automatically make the world a safer place.

Violent movements that pay no heed to international law, and conflicts between states that should be cooperating in the Security Council, have even made the international legal order less effective in recent years. Some states consciously undermine the legal force of judgements of the International Court of Justice and the jurisdiction of the International Criminal Court. The instability of international relations therefore extends to the international legal order itself.

In contrast to the traditional security policy, greater account needs to be taken of the violence committed by entities other than states. Groups like Da'esh have no respect whatsoever for the norms of international humanitarian law concerning methods of warfare and the treatment of prisoners of war as laid down in treaties. The sovereignty of states over their own territory does not impede these organisations, but rather offers them an absurd form of protection. For example, it initially deterred the Netherlands from participating in international operations against Da'esh in Syria. There are many countries where terrorist activities can be prepared and weapons can be bought and the associated financial transactions can be carried out. Some of those countries are called ‘fragile’ states. The fragility of the system of international relations is encouraged by the fragility of states, a fragility that in turn undermines stability in states that previously felt secure.

However, the problem is not just with states in which the power of the government is waning or has actually disappeared. There are also states without internal controls. States like North Korea, which is developing offensive nuclear weapons, are not inhibited by internal checks and balances. Even ‘strong’ states can evoke internal and external tensions and generate violence[10]; the Russian Federation is increasingly an example of such a state. The military capacity to withstand intimidation by such states is the first step in the undoubtedly lengthy process of restoring the international legal order as the basis of mutual trust.

In that respect, finally, it is important to realise that, in one way or another, every regime and every conflict arises from and is a response to how people experience their lives. Moïsi has referred to the role played by emotions such as fear, humiliation and hope in international relations.[11] Repressive regimes are normally the product of disrupted social relations (which might explain them, but naturally can in no way justify them). Warmongers repeatedly invoke the fact that they are defending their ‘own’ people, group or religion, which they argue deserves to prevail over the enemy. The politics of peace has to take account of that. Improving living standards in the realisation that the world is held together by transnational relationships lays the basis for that, but is not enough. The resilience of free societies and the enforceability of fundamental precepts of international humanitarian law have to be strengthened at the same time.

Strengthening defence is therefore not a new form of entrenchment, but an acknowledgement of the demands that an interconnected world also impose on the Netherlands.

7.3 Conclusions and Recommendations

On the basis of the analysis in this book we arrive at three main conclusions, together with related recommendations:

1. The security environment in which the Netherlands finds itself has deteriorated. Organise the security policy and apparatus accordingly.
2. The Netherlands’ defence policy should be guided by the obligations arising from the constitution and the country’s alliances (NATO, EU).
3. The decline in the sustainability of the armed forces and the deterioration in the security environment demand a tighter focus and additional investment.

These conclusions are explained and translated into recommendations below.

7.3.1 First Conclusion

The security environment in which the Netherlands finds itself has deteriorated. Organise the security policy and apparatus accordingly.

In light of the developments in recent years, there can now be no doubt that the security environment in which the Netherlands finds itself has deteriorated. The shooting down of Flight MH17 in the airspace over Ukraine and the arrival of larger numbers of refugees from Syria and other countries underline the fact that a lot of the political turbulence and the conflict zones elsewhere in the world also affect the security of our country, directly or indirectly. The period of Russian membership of the Council of Europe and its guarantees of respect for the rule of law were followed within just a few years by a growing alienation, which runs so deep that the country's military capacity (including nuclear weapons) is once again seen as a threat. With the political transition in the US, from 2017 Europe will probably have to increasingly rely on itself to maintain a credible military capability. At the same time, however, the strength of the only recently adopted Common Foreign and Security Policy, as laid down in Title V of the Treaty on European Union, will be impaired by the departure of one of the Union's most important member states and by the waning support for the Union's shared values in some recently admitted member states. The prospects of stabilising or expanding Turkey's engagement with the EU have also been seriously harmed by the development of that country's political system into an autocracy. Instability and the impairment of fundamental legal principles in and around the continent of Europe are causing insecurity and compromising international peace. There are many fragile states in Africa and the Middle East, where unrest and conflict will have serious implications for Europe, for example in the form of migration. Furthermore, the turbulence is unlikely to abate in the foreseeable future. The EU's capacity to act and the allied security at NATO level, and hence also the capabilities and the efforts of the Netherlands in the long term, will therefore also be put to the test.

The Netherlands is vulnerable to the negative manifestations of these developments. Nationally because of the negative transnational consequences of globalisation; regionally because of the extensive economic and political interconnectedness of the EU; and globally because of the strain on the rule-based international order, from which the Netherlands has benefited so greatly in recent decades and which greatly determines its stability and prosperity.

In its report *Attached to the World* (2010), the WRR observed that the Netherlands needs a strategic foreign and security policy that takes account of the changing environment and its own priorities. Such a policy goes beyond formulating general intentions and orientations. The Netherlands must act more effectively at the interface of the worlds of geopolitics and networks, set clearer priorities and explore where it can make a difference amongst the numerous threats, risks and opportunities on the European and global chess board.[12] It is no longer possible to assess those threats, risks and opportunities from a complacent Dutch perspective. International peace is an indivisible, but no longer self-evident common good of the Kingdom of the Netherlands, its allies and its surroundings.

Recommendation 1. Strengthen the connection between internal (national) and external (international) security policy.

An important element of a security policy tailored to the altered environment is the relationship between national and international (internal and external) security. That conclusion has already been reached in numerous reports and studies, including the Clingendael Institute's Strategic Monitor in 2014, which included a wide-ranging discussion of the relationship between internal and external security (immigration, terrorism, transnational crime and cyberspace). The conclusion was that "the interconnectedness of external and internal security is a fact. Structural cooperation between actors with responsibility for security 'inside' and 'outside' the Netherlands is failing to keep pace with that fact. That calls for an integrated security strategy and intensification of the cooperation between external and internal security actors".[13] Internal security and external security are intertwined.

That plea has, if possible, become even more relevant. Furthermore, even now there is still no coherent approach in those areas where the Netherlands can really make a difference. As Chapter 5 showed, in themselves the three main tasks of defence policy remain the same. The difference is that they give rise to other and more specific challenges and the main tasks must be consistently interpreted in the context of the interconnectedness of internal and external security. That calls for a revised and more coherent strategy. The substantive coherence can be enhanced on the basis of the orientation to human security, national security and flow security.

This book recommends placing greater emphasis on the relationship between internal and external security. Specifically, that can be accomplished by strengthening the Royal Netherlands Marechaussee and making greater use of it in both contexts and reflecting the relationship between internal (national) and external (international) security at ministerial level by assigning a greater coordinating role to the prime minister.

Recommendation 2. Merge the two security strategies (National Security Strategy and International Security Strategy).

Despite efforts to modernise it, Dutch security policy is not geared to the changes in the security environment. Security policy is still highly compartmentalised and 'traditional', with internal and external security treated as separate domains. That is not a barrier to operational cooperation in specific policy areas, but there is still no question of an integrated approach. The same applies equally for other policy areas. For example, the policy towards fragile states is laid down in various policy documents and spread over three budgets. The EU plays a major role in reducing instability in Africa, but the efforts under the Common Security and Defence Policy, the external dimension of the Justice and Home Affairs policy (border control, illegal immigration and repatriation policy) and the European Neighbourhood Policy call for an integrated, coherent strategic vision document setting out how the Netherlands intends to tackle instability in Africa and its underlying causes.

Our recommendation is to merge the National Security Strategy and the International Security Strategy, while retaining the system laid down in the National Security Strategy. There is already considerable overlap between the two strategies. An important reason for integrating the two strategies is that the relationship between internal and external security will be established from the outset of the strategy-formulation process. The Netherlands' security strategy – and the strategy for the national armed forces derived from it – must be explicitly embedded in the security strategies of the EU and NATO. That does not mean simply following them. It is legitimate to incorporate insights and preferences that match the country's own interests and capabilities in the strategy.

Recommendation 3. Establish a Netherlands Security Council to formulate the security strategy.

The structure for formulating the envisaged security strategy can be created by establishing a Netherlands Security Council by law pursuant to Article 79 of the Constitution.[14] The Council would be chaired by the prime minister and also include the other relevant ministers and the most senior civilian and military civil servants, as well as experts chosen for their professional authority. A cabinet subcommittee, such as the existing Council for the Intelligence and Security Services (RIV), would, together with the Ministerial Security Committee (MCV), perform the important role of coordinating the sharing of information, policy and consultation between politicians and officials in the area of security. A Netherlands Security Council would also perform other supplementary functions in relation to formulating strategy. In the first place, such a Council would guarantee the alignment of domestic and foreign policy across the entire spectrum of security policy. Secondly, the Security Council would be directly linked (with the assistance of a Security Planning and Research Agency) to research institutes and the academic world to provide impulses for the strategic security policy. Thirdly, a Netherlands Security Council would be the forum that visibly reflects the relationship between internal and external security. The importance of a Netherlands Security Council is not reserved to large countries like the United States. Smaller countries like the Netherlands can also not automatically count on a fairly transparent landscape or assume that issues can be arranged decentrally, according to a careful (interdepartmental) balance of power or via individual politicians. Particularly small developed countries, which by definition face an overfull internal and external security agenda, have to excel in strategy formulation. There are no constitutional objections to establishing a Netherlands Security Council. The Netherlands Security Council would have to reflect the socialisation and expansion of the concept of security in its composition, while also embodying the relationship between internal and external security in its structure. The Netherlands Security Council would explicitly not be solely concerned with national security, but would also be oriented towards the possibilities and goals of international and EU cooperation and coordination. The Netherlands Security Council should not depend entirely on analyses by the intelligence and security services. A multidisciplinary Security Planning and

Research Agency should support the strategy formulation process with a continuous supply of thorough analyses.

Recommendation 4. Establish a Security Planning and Research Agency to support strategy formulation by the Netherlands Security Council.

There should be a Security Planning and Research Agency to support the Netherlands Security Council and which also has its own tasks in terms of conducting research and producing regular reports, strategic analyses and scenarios. A Security Planning and Research Agency would prepare the consolidation of the two security strategies (the National Security Strategy and the International Security Strategy) and draft the new strategic document, which would be fleshed out by the Netherlands Security Council. In addition to establishing and developing its own expertise to support the Netherlands Security Council, the Security Planning and Research Agency should explicitly serve as a network model in bringing together and making maximum use of the extensive knowledge that already exists. The Security Planning and Research Agency would not be a substitute for the existing knowledge institutes, for strategy formulation within the ministries or for periodic reports such as the Security Monitor, but would supplement them. The Security Planning and Research Agency should look for connections with the knowledge agenda and the science agenda in the Netherlands and seek to engage in the public debate. It seems reasonable that the Security Planning and Research Agency would fall under the direction of the Ministry of General Affairs.

Recommendation 5. Invest more in the knowledge function and strategic thinking within the armed forces and elsewhere.

The Netherlands needs a clearer – and more focused – *strategic* vision and needs to do more in terms of vision development. Additional budgetary scope is needed to align the capabilities of the armed forces with the specified ambitions. Above all, choices will have to be made in the priorities (and the cohesion between them) in relation to the different branches of the armed forces. Investments will only be able to achieve their goals if there is a transparent, convincing and widely supported strategy in place. The strategy development process is now too often a poor relation. 'Strategic illiteracy',[15] 'strategic illiteracy' and 'strategic dyslexia'[16] are some of the unflattering and very worrying descriptions used to describe the status of strategic thinking in the Netherlands. A long-term investment in the development of a strategic vision is therefore badly needed. Special attention should be devoted to the preventive phase and to flow, cyber, human and border security and the geopolitics of emotions.

The establishment of the Netherlands Security Council and the Security Planning and Research Agency should provide impulses by making research budgets available. Another recommendation is to expand academic research and education in these fields. The link with the National Science Agenda could provide an important boost. There should also be a heavy emphasis on interdisciplinary research that can enhance analysis and strategy formulation and improve the availability and interpretation of data.

Recommendation 6. Strengthen diplomacy and the policy of diplomatic missions to increase anticipatory capacity.

Crisis management and conflict prevention by improving the living conditions in fragile states is not just a question of respecting human rights and thereby promoting the international legal order (Article 90 of the Constitution). In an highly interdependent world, they are also a matter of enlightened common interest. What happens in those states has an impact on the situation here. The integrated approach to security and development must therefore be maintained, despite the not universally positive experiences with that approach in Uruzgan.

Security requires anticipation and foresight in the relevant countries and regions. Whereas until the 1990s military threats dominated the security discourse, in the last two decades development, economic and ecological issues have risen higher on the agenda and are increasingly assuming the aspect of a security policy issue. Partly in light of that, international security will have to be linked to the Sustainable Development Goals, which will also alter the geographic scale of security issues. Regional and global interrelationships and transnational issues will demand investment in development, collective security and supranational partnerships.

The Netherlands is a member of networks which it can use to increase the sensitivity for international security. The network of diplomatic missions has more than an economic function. The presence of justice, police and defence attachés should therefore also be increased, and not just in the foreign missions. Strengthening the network is only purposeful if the coherence and exchange of information is continued within the ministries themselves and the findings make their way into the strategic preparation of security policy. The Netherlands Security Council and the Security Planning and Research Agency could also play a role in this regard.

Security interests also call for extra investment in development cooperation. The role of development cooperation is crucial. The budgetary flexibility that the WRR advocated in *Minder pretentie, meer ambitie* [Less pretention, more ambition] (2010) should not be understood solely in a downward sense, but can also move in an upward direction. If the situation requires it, as it does now, extra investment in the armed forces ('Defence') should go hand in hand with the strengthening of development cooperation and trade relations ('Development') and international diplomacy ('Diplomacy') in accordance with the 3D doctrine. In light of the deteriorating security environment and the need for a coherent strategic policy, the target of spending 0.7% of GDP on development cooperation as agreed in the UN should again be the guiding principle. Otherwise the ambitions for genuinely integrated and interdisciplinary strategic operations will perish.

The government should formulate further proposals for intensifying cooperation and for joint efforts to address identified European military shortfalls and to strengthen European diplomatic actions and development policy in consultation with strategic partners. This should also enhance the strategic function, the knowledge and anticipatory capacity, for the purposes of the national security strategy. The defence planning process could also be placed in an interdepartmental context in order to integrate the armed forces as effectively as possible in a comprehensive approach to security.

7.3.2 *Second Conclusion*

The Netherlands' defence policy must be genuinely guided by the obligations arising from the constitution and the country's alliances (NATO, EU).

With the fall of the Berlin Wall and the ensuing détente, the need to contribute to NATO was felt less keenly. The direct threat disappeared and the armed forces could work with other organisations in contributing to stability elsewhere. That view of the world has been under pressure for some time. On the one hand, on both sides of the ocean separating the NATO countries there are tendencies that put national interests first and push multilateralism into the background. On the other hand, the direct threat has returned. The risk that as a result defence policy will become inward-looking is not imaginary.

But it is precisely the changed security environment, in combination with recognition of the Netherlands' modest position and role in a hardened world, that necessitates international cooperation. The existing alliances are of fundamental value at the present time. Without these ties, the position of small countries is extremely uncertain. Both citizens and the government should respect the implicit and explicit requirements of the social contract, which include providing security. Otherwise, mutual mistrust will dominate. What applies for citizens and government, also applies between governments, or between the government and international organisations like NATO and the EU.

Recommendation 7. Invest primarily in the NATO alliance.

NATO is also the Netherlands' most robust alliance. The role of the Netherlands in NATO is laid down by treaty as provided for in the Constitution and comprises contributions to exercises and operations, investments and strategy formulation. NATO is of fundamental importance to the Netherlands. The Netherlands is itself an important actor and makes a real contribution to essential operations. However, the Netherlands has failed to meet its financial obligations. This is putting a strain on mutual trust, not to mention the reduction of capabilities that it leads to. The recommendation is that the Netherlands promotes the alliance in every respect, particularly financially, by means of contributions and investments. This also means that the Netherlands must take priorities in NATO planning seriously and must contribute to the elimination of identified shortfalls.

Recommendation 8. Invest in cooperation within Europe on the basis of the existing relationships.

The emphasis in Europe must be on cooperation in investment, the choice of materiel and operationalisation, not necessarily on the creation of a 'European defence force'. Calls for a 'European army' are not only a pipedream, they also distract attention from the successful partnerships that already exist and are needed. Examples are the partnerships with Belgium, Germany, France, the United Kingdom and Norway. The international cooperation also extends to the procurement and maintenance of materiel. It also involves joint deployment without any impairment of sovereignty. This promotes the efficiency and effectiveness of any deployment of the army. These connections are important and should receive greater emphasis.

Cooperation also strengthens defence capabilities. The Netherlands benefits from a close relationship with Germany as the repositioning of the German armed forces within Europe takes further shape. The existing partnerships could be strengthened by making use of permanent structured cooperation on the basis of Article 42(6) and Protocol no. 10 of the Treaty on European Union, which creates the possibility for particular member states to intensify their cooperation in military affairs. It also means that the Netherlands must make greater efforts to eliminate identified shortfalls in Europe.

The integrated approach of security and development at EU level should therefore be strengthened. In the so-called 'belt of instability' around the EU, where the EU and the Russian Federation and where the EU and Arab countries are close neighbours, the European Neighbourhood Policy serves common interests. The Netherlands should press for substantial investment in the Eastern Partnership and the Union for the Mediterranean. Initiatives such as the European Defence Action Plan (EDAP) should also receive more attention.

This book stresses that investments should always be accompanied by further investigation and expansion of the possibilities for national and international military cooperation. Military cooperation within NATO and the EU should be the point of departure when making choices about investments. That does not mean that additional specific investments in the armed forces are not required. On the contrary. But the framework for setting priorities is cooperation and the joint performance of tasks, whether within NATO or the EU. This book also recommends continuing and expanding these partnerships, even when investment increases and there is a larger budget (in other words, even if there is no immediate financial necessity).

7.3.3 Third Conclusion

The decline in the sustainability of the armed forces and the need for more active anticipation of security risks demand a tighter focus and additional investment.

To appreciate the importance of a strategic security policy, it is essential to understand that the concept of security and the security policy agenda have changed fundamentally twice in the last quarter of a century. After the fall of the Berlin Wall, the direct military threat disappeared and the emphasis came to lie on developing rapid reaction intervention capacity, based on a narrative of preserving the status quo. This policy has been steadily losing relevance and conviction recently, as the environment in and around the European continent has proved unexpectedly turbulent. There have also been violent swings in expectations: just a few years after the rejoicing at the Arab Spring, the dominant picture is one of horrific internal wars and the unbridled violence of Da'esh. The classical nineteenth-century notion of protecting state or territorial security is still a key objective of the government's security policy, but the threats to freedom and security have also assumed entirely different guises than that of occupation of the territory by hostile powers. In the current situation, protecting connections that are essential for society and preventing terrorist attacks, cyber attacks and other forms of aggressive disruption demand special attention.

Recommendation 9. Raise defence spending to 2% of GDP in predetermined increments, adhere to the NATO guidelines for investments and draw up a long-term plan that also embraces development and diplomatic prevention.

The armed forces must also remain capable of performing their three statutory tasks (defending the territory of the Netherlands and its NATO allies, promoting the international legal order and supporting the civil authorities in the Netherlands and other countries) in a changing security situation. To that end, first and foremost it is necessary to eliminate the existing constraints on their operational deployability. However, that is by no means the whole story. In light of the changed security situation, additional efforts are required in relation to all three domains, in terms of defending NATO territory, preventing instability around Europe and supporting civil authorities, in particular with regard to crisis management in the context of national security. There are three steps that need to be taken.

The first step is to increase the sustainability of the armed forces in the short term. This is dependent on the recuperation of the armed forces, on the basis of which steps can be taken to further strengthen them. Earlier, in the policy document *In the interest of the Netherlands* (2013), the government opted for a widely deployable armed forces, which, in light of the available budgetary frameworks at the time, forced a second choice: a reduction of the sustainability of the armed forces. In light of the changed security situation, which imposes greater demands on the state of readiness and preparedness of the armed forces, the recuperation and sustainability of the armed forces deserve every attention.

The second step involves strengthening the Dutch armed forces with a view to improving the EU's capacity to act and NATO's deterrence and defence capabilities, partly against the background of burden-sharing with the US. This is a longer-term perspective, with intensification of cooperation with strategic partners and – joint – reduction of the identified military shortfalls in Europe as the points of departure. The Dutch armed forces are in an excellent starting position in this regard, with extensive cooperation with Germany and Belgium as the basis.

The third step concerns the perspective for the Dutch armed forces in the longer term. The changed security situation and, more generally, the steadily accelerating pace of technological and social changes, call for innovations in the operational, doctrinaire, personnel and technological domains. Replacing and modernising platforms does not provide sufficient certainty of being able to keep pace with technological and operational developments over time. The complex security environment calls for the development of new systems, in close consultation with research institutes and private actors. Innovation should be a continuous process, but the long-term perspective is important. Depending on the option that is chosen, this could, for example, mean that the Netherlands primarily makes a far greater contribution to naval capacity and in conjunction with that enhances the training and economic position of the maritime sector.

Under the NATO agreements on raising defence spending that were made in Wales in 2014 and endorsed in Warsaw in 2016, the government cannot avoid earmarking additional funding for defence in the coming years. The importance of that for a long-term perspective cannot be stressed enough. The absorption capacity of the armed forces brings with it the need to establish a timetable for extra investments and for creating focus. A multi-year budget provides certainty and stability for that process. The discussion does not have to lead to a final decision on whether or not a multi-year budget should coincide with a government period (in other words, either 4 years or 5 or 6 years, for example). But from the time the new government takes office, the target of 2% will have to be reached in 2024, 10 years after the agreements in Wales (2014), in predetermined increments.

Security policy extends beyond defence policy alone. Budget increases to make the improvements should therefore not only go to defence policy, but also include targeted investments in foreign policy, and development cooperation in particular.

Recommendation 10. Improve the coherence of policy towards the new (hybrid) challenges in relation to border security, cyber security, flow security and human security.

The expansion and socialisation of the security agenda and the increased interconnectedness of internal and external security are reflected in Dutch security and defence policy. Greater attention is devoted to human security, border security, flow security and cyber security. The importance of flow security for the Netherlands must not be underestimated and policies relating to energy, climate, raw materials and the cyber domain should therefore be incorporated in security policy. The focus on crisis and stabilisation missions also contributed to the calls for an integrated approach to security and development issues. The rise of national security as an area of attention resulted in the drafting of separate strategic documents for national and international security policy, as well as specific strategies for counterterrorism and cyber security. Our recommendation is to strengthen the coherence between these domains and intensify the efforts.

Recommendation 11. Continue with the existing specialisation, with the addition of capabilities for cyber warfare.

For a long time the allocation of resources to the different branches of the armed forces was a routine This book's recommendation is to significantly tighten the focus and introduce specific investments in the process. When it comes to future investments, continue the existing specialisation with partners and by branches of the armed forces, using the choices that have already been made for capital investments in the air force and the navy as the point of departure. Protecting Dutch vessels has traditionally been a task of the navy. As a seafaring trading nation, the Netherlands has a particular interest in secure shipping routes. The navy also plays an important role in providing security in the Caribbean parts of the Kingdom.

These interests and capabilities must not be neglected in the choice of future investments and the coordination of priorities within NATO and the EU. But the deployment of the navy should not be seen in the context of a traditional division of 'land, air and sea' roles; the operational coherence with the air force, for example, is essential. Thinking in terms of effects and capabilities is far more important than clinging to existing divisions of roles. It will also provide the flexibility to respond to identified shortfalls at NATO and EU level. Strengthening the Marechaussee is also important for missions to support the police and judiciary and European border control, as well as operations in the Caribbean region.

The capabilities in relation to cyber warfare demand a place of their own, especially in light of the escalation on the part of Russia. In 2016, NATO recognised cyberspace as a fourth domain in addition to land, air and sea. That could mean, but does not necessarily have to mean, that there will have to be a separate branch of the armed forces for that domain. The most important thing, particularly in light of the interconnectedness of cyber in all of the branches, is recognition of the connection. But the subject does require extra attention and effort. The existing Defence Cyber Command, whose structure is currently based on the traditional branches of the armed forces, will have to evolve into an organisation that is prepared for far greater risks.

Recommendation 12. Continue the socialisation of the armed forces and strengthen the support of civil authorities and the civil-military cooperation within a balanced and transparent security culture.

Governments will be confronted more frequently with transnational (security) issues whose origin, course and consequences are difficult to oversee. In a modern democracy, this area of tension is under the microscope of the media, critical NGOs and the articulate general public. At the same time, complex, highly developed and open societies are relatively vulnerable to groups and individuals that take advantage of the blurring of the boundaries between 'internal' and 'external'. Furthermore, the public often has little tolerance for risks and uncertainties, precisely because of the high expectations for government intervention. A government that fails to meet those expectations with visible security measures can quickly engender feelings of fear and insecurity. But a government that does meet them – although a complete guarantee of security is of course impossible to give – can actually reinforce feelings of insecurity. The excessive politicisation of security policy in society can lead to a negative spiral of fear and mistrust, thus endangering the open society itself.

In addition to the task of averting direct threats (averting terrorist attacks, the use of the Royal Netherlands Marechaussee for border control and the deployment of soldiers for security stand out) governments, international organisations and other actors in society have the task of anticipating potential and even imaginary future threats and risks, for example by taking precautionary measures. The recommendation is therefore to continue the socialisation of security policy, on the basis of an expansion of that policy, but to guard against (and provide guarantees to prevent) an all-pervasive security culture.

Endnotes

1 Article 51.

2 Crawford, J. (2012: 130–131).

3 Article 1(1) of the International Covenant on Civil and Political Rights and the identical Article 1(1) of the International Covenant on Economic, Social and Cultural Rights. Both treaties were concluded on 16 December 1966.

4 Hasenclever, A. (2014: 439).

5 Bourbeau, Ph. (2015: 11).

6 Guzzini, S. (2012).

7 Ferdowsi, M.A. (2002).

8 Olmer, J. (2016, chapter 5).

9 The development cooperation policy and the security policy can then be linked to each other in a more useful manner than with the current Official Development Assistance (ODA) allocations.

10 Crawford, J. (2012: 128).

11 Moïsi, D. (2009).

12 WRR. (2010a, b: 9–10; 29; 85).

13 Rood, J. (2014: 130–131).

14 By virtue of the third paragraph, duties in addition to advisory ones may be assisgned to permanent advisory bodies by or pursuant to Act of Parliament.

15 Duyvesteyn, I.G.B.M. (2013).

16 Amersfoort, H. (2016: 6).

Appendix I: List of Persons Consulted

Name – Organisation

E. Akerboom – Ministry of Defence
T. Akesson – Ministry of Defence, Sweden
R. Allison – Oxford University
H. Andersson – Ministry of Defence, Sweden
J. Andriessen – Ministry of Foreign Affairs
Anthony – Stockholm International Peace Research Institute
Armin Staigis – Bundesakademie für Sicherheitspolitik (BAKS)
K. Ash – OECD Paris
Bakker – Ministry of Foreign Affairs
W. Baron – Netherlands embassy in Berlin
H-P. Bartels – German Bundestag
K. Bartels – Military Committee
J. Bartholdson – Folk och Försvar, Sweden
J.W. Beaujean – Netherlands embassy in Paris
M. van Beijnum – Clingendael Institute
R. Bekink – Netherlands embassy in Washington
M. Bergquist – Royal Swedish Academy of War Sciences
B-G. Bergstand – FOI Swedish Defence Research Agency
R. Beste
H. Binnendijk – School of Advanced International Studies, United States
J. Blacquière – Netherlands embassy in Paris
M. Boef – Ministry of Foreign Affairs
M. de Boer – Adviser on Uruzgan & Embassy in Kabul
J. Brandt – Ministry of Foreign Affairs
H. Brauss – Defence Policy and Planning Division, NATO
S. Brimley – Center for a New American Security, United States
Th.F.M. Brinkel – Netherlands Defence Academy

E. Hirsch Ballin et al. (eds.), *Security in an Interconnected World*,
Research for Policy, https://doi.org/10.1007/978-3-030-37606-2

J. Broeks – Permanent Military Representative, NATO
M.J. Burrows – Atlantic Council, United States
F. Burwell – Atlantic Council, United States
J. Cederberg – Department for Security Policy
M. Chalmers – Royal United Services Institute
F. Chantereau – Department for Security Policy
Chiu – Atlantic Council, United States
Chollet – The German Marshall Fund of the United States
K. Colijn – Clingendael Institute
J. Corbett – Foreign and Commonwealth Institute, United Kingdom
Cushman – National Defense University, United States
Damen – Political and Security Committee (PSC)
M. Dickow – Stiftung Wissenschaft und Politik, Germany
W. van Dijk – Cyber Security/National Coordinator for Security and Counterterrorism (NCTV)
M. Drent – Clingendael Institute
M. du Mont – Department of Defense, United States
P. Ducheine – Netherlands Defence Academy/University of Amsterdam
Duivesteijn – University of Leiden, University of Utrecht
J. van Elk – Ministry of Foreign Affairs
P. Engelke – Atlantic Council, United States
C.H. de la Faverie du Ché – Ministère des Armées, France
de Feijter – Political and Security Committee (PSC)
S. Feyouk – TBC
J.P. Filiu – Sciences Po
S. Filippini – Cordaid
Fisher – National Defense University, United States
Frerks – Netherlands Defence Academy
W. Geerts – Ministry of Defence
P. Gelton – National Coordinator for Security and Counterterrorism (NCTV)
de Graaf – University of Utrecht
Grand – Fondation pour la Recherche stratégique, France
N. Granholm – FOI Swedish Defence Research Agency
F.C. Günther – Büro des Wehrbeauftragten, Germany
Gyllensporre – Ministry of Defence, Sweden
Hamilton – Johns Hopkins University, School of Advanced International Studies (SAIS), United States
Ph. de Heer – Ministry of Foreign Affairs
Hellema – University of Utrecht
M. Holmström – Dagens Nyheter, Sweden
J. W. Holtslag – Former Secretary-General, Ministry of the Interior and Kingdom Relations
J.W. Honig – Cambridge University, King's College
Hoogstraten – Ministry of Defence

J.G. de Hoop Scheffer – Advisory Council on International Affairs (AIV)/University of Leiden
N. van Hulst – OECD Paris
N. Jaarsma – Ministry of Foreign Affairs/Security Policy Department (DVB)
M. Jadoul – Ministry of Foreign Affairs
Jan – Ministry of Defence
R. Jones-Bos – Ministry of Foreign Affairs
J. de Jonge – TNO Innovation for Life
T. Joustra – Dutch Safety Board
D. Jurt – Ministry of Defence, Switzerland
J. Kenny – European External Action Service (EEAS)
D. Keohane – FRIDE
W. Kolbow
D. Kooij – Dutch Embassy in Berlin
T. Koster – Plans and Policy Division, NATO
Kronenburg – Ambassador in Paris
M. de Kwaasteniet – Ministry of Foreign Affairs
M. Kwast – Advisory Council on International Affairs (AIV)
J. Lagerlöf – Department for Security Policy, Sweden
J. van der Lijn – Stockholm International Peace Research Institute
R. Lindsay – Foreign and Commonwealth Office (FCO), United Kingdom
D. Lonete – DG DEVCO
Lord – Department of Defense, United States
S. Loughhead – Department for International Development, United Kingdom
Luedtke – Ministry of Defence, Sweden
M. Lutje Schipholt – Allied Command Transformation (ACT)
Magnusson – Royal Swedish Academy of War Sciences
J. Malminen – FOI Swedish Defence Research Agency
U. Mans – University of Leiden
McGee – Department for Strategy and Planning
Mennen – National Institute for Public Health and the Environment (RIVM)
Merritt – SDA Friends of Europe
Molenaar – European External Action Service (EEAS)
F. Mollen – Netherlands Embassy in Berlin
G.F. Mouton – Secrétariat de la Défense et de la Sécrurité nationale (SGDSN)
M. Nelson – Stanford University
M. O'Hanlon – Brookings Institution, United States
F. Osinga – Ministry of Defence
Peersman – Permanent Representation of the Kingdom of the Netherlands to the European Union
M. Popowski – European External Action Service (EEAS)
F.P. van Putten – Clingendael Institute
R. Querido – Permanent Representation of the Kingdom of the Netherlands to the European Union
Ract-Madoux – Gouverneur des Invalides

J. Radisch – OECD
R. Reefman – Ministry of Defence
S. Reijn – Ministry of Defence
J. Rheinberg – Foreign and Commonwealth Office (FCO), United Kingdom
Riecke – TBC
Rihan- Cypel – La commission de la défense nationale et des forces armées, Assemblée nationale
van Rijnsoever – Permanent Representation of the Kingdom of the Netherlands to the European Union, Political and Security Committee
Rood – Clingendael Institute
H. Rouw – Peace organisation PAX
P. de Ruiter – Allied Command Transformation (ACT)
Rutgersson – Department for Security Policy
A.E. Sekowski – Department of Defense, United States
Shea – Deputy Assistant Secretary General for Emerging Security Challenges, NATO
Sie Dhian Ho – Clingendael Institute, Advisory Council on International Affairs (AIV)
S. van der Sluis – Ministry of Foreign Affairs
S. De Spiegeleire – The Hague Centre for Strategic Studies
T. Stamatopoulos – Ambassador in Brussels
K. van der Steenhoven – Ministry of the Interior and Kingdom Relations
Stelzenmüller – Brookings Institution, United States
Svoboda – Overseas Development Institute, United Kingdom
T. Sweijs – The Hague Centre for Strategic Studies
J. Techau – Carnegie Europe
Tuominen – Defence Attachés
van Tuyll – National Coordinator for Security and Counterterrorism (NCTV)
J. Twiss Quarles van Ufford – Ministry of Foreign Affairs
M. Urlings – Advisory Council on International Affairs (AIV)
J. Vaïsse – Centre d'Analyse, de Prévision et de Stratégie (CAPS), France
Vershbow – Ambassador in Brussels
J. Voorhoeve – Advisory Council on International Affairs (AIV)
J. van Vugt – National Coordinator for Security and Counterterrorism (NCTV)
J. Wagner – Civil Military Cooperation Centre of Excellence
Whaites – OECD
J. Wiers – Ministry of Foreign Affairs
R. de Wijk – The Hague Centre for Strategic Studies
J. Wijnands – Ministry of Foreign Affairs
J. de Wilde – University of Groningen
M. Winnerstig – Defence Analysis
Wyatt – Department for Strategy and Planning
Th. Zaalberg – Netherland Institute of Military History
Zandee – Clingendael Institute
Zwaanswijk – Defence attaché Sweden/Denmark

Appendix II: List of Acronyms and Abbreviations

A2/AD	Anti-access/Area Denial
ACT	Allied Command Transformation
AGS	Air Ground Surveillance
AIV	Advisory Council on International Affairs
AIVD	General Intelligence and Security Service
ARV	Netherlands Security Council
GDP	Gross Domestic Product
BENESAM	Belgian-Dutch Naval Cooperation
BIV	Budget for International Security
BRICS countries	Brazil, Russia, India, China and South Africa
BZ	Foreign Affairs
BZK	Interior and Kingdom Relations
CDA	Christen Democratisch Appèl
CEPS	Centre for European Policy Studies
CSBA	Center for Strategic and Budgetary Assessments
Da'esh	Al-Dawla al-Islamiya fil Iraq wa al-Sham
DGPZ	Director-General for Political Affairs
EDAP	European Defence Action Plan
EEAS	European External Action Service
EPSC	European Political Strategy Centre
EU	European Union
CSDP	Common Security and Defence Policy
HCSS	The Hague Centre for Strategic Studies
HIV/AIDS	Human Immunodeficiency Virus/Acquired Immunodeficiency Syndrome
IBO	Interdepartmental Policy Study
ICMS	Enhanced Civil-Military Cooperation
ICT	Information and Communications Technology
IIJ	The International Institute for Justice and the Rule of Law
IOB	Inspectorate of Development Cooperation and Policy Evaluation

E. Hirsch Ballin et al. (eds.), *Security in an Interconnected World*,
Research for Policy, https://doi.org/10.1007/978-3-030-37606-2

IOM	International Organization for Migration
ISAF	International Security Assistance Force
ISS	International Security Strategy
JBZ	Justice and Home Affairs
JEF	Joint Expeditionary Force
KNP	Commissioner of the Netherlands Police
KPN	Koninklijke PTT Nederland N.V.
LCF	Air-Defence and Command Frigates
MALE UAV	Medium-Altitude Long-Endurance Unmanned Aerial Vehicle
MCV	Ministerial Security Committee
MENA	Middle East and North Africa
MINUSMA	Multidimensional Integrated Stabilization Mission in Mali
MIVD	Military Intelligence and Security Service
MJSP	Multi-year Strategic Plans
NATO	North Atlantic Treaty Organization
NCC	National Crisis Centre
NCTV	National Coordinator for Security and Counterterrorism
NGO	Non-governmental organisation
NRB	National Risk Assessment
NRF	NATO Response Force
NSOB	Netherlands School of Public Administration
NSS	National Security Strategy
NVP	National Risk Profile
ODA	Official Development Assistance
OECD	Organisation for Economic Co-operation and Development
OM	Public Prosecution Service
OPV	Ocean-Going Patrol Vessels
OSCE	Organization for Security and Co-operation in Europe
PESCO	Permanent Structured Cooperation
PBL	Netherlands Environmental Assessment Agency
R2P	Responsibility to Protect
RAP	Readiness Action Plan
RIV	Council for the Intelligence and Security Services
RIVM	National Institute for Public Health and the Environment
RLI	Council for the Environment and Infrastructure
SACEUR	Supreme Allied Commander Europe
SDGS	Sustainable Development Goals
SIPRI	Stockholm International Peace Research Institute
SNV	National Security Strategy
SGP	Staatkundig Gereformeerde Partij
TEU	Treaty on European Union
UNDP	United Nations Development Programme
USD	United States Dollars
V&J	Justice and Security
VCMS	Enhancing Civil-Military Cooperation

VJTF	Very High Readiness Joint Task Force
UK	United Kingdom
UN	United Nations
VNO/NCW	Confederation of Netherlands Industry and Employers
US	United States
VVD	Volkspartij voor Vrijheid en Democratie
TFEU	Treaty on the Functioning of the European Union
WGA	Whole-of-Government Approach
WHO	World Health Organization
WODC	Research and Documentation Centre of the Ministry of Justice and Security
WRR	(Wetenschappelijke Raad voor het Regeringsbeleid) Scientific Council for Government Policy)
WTO	World Trade Organization

Bibliography

Abels, P. H. A. M., & Willemse, R. (2004). Veiligheidsdienst in verandering; de BVD/AIVD sinds het einde van de Koude Oorlog. *Justitiële Verkenningen, 3*, 83–98.

Abrahamsen, R., & Williams, M. (2010). *Security beyond the state. Private security in international politics*. Cambridge: Cambridge University Press.

Abrahamson, H. (2008, April 7–9). Understanding world order, continuity and change. security, development and justice in the global era. *Some preliminary reflections for the China – Nordic Peace Research Conference*, Beijing.

Adelman, M. A. (1972). *World petroleum market*. Baltimore: Johns Hopkins University Press.

Adviesraad Internationale Vraagstukken (AIV). (2000). *Humanitaire interventie* (Vol. 13). The Hague: AIV.

Adviesraad Internationale Vraagstukken (AIV). (2004). *Falende staten. Een wereldwijde verantwoordelijkheid* (Vol. 35). The Hague: AIV.

Adviesraad Internationale Vraagstukken (AIV). (2009). *Crisisbeheersingsoperaties in fragiele staten. De noodzaak van een samenhangende aanpak* (Vol. 64). The Hague: AIV.

Adviesraad Internationale Vraagstukken (AIV). (2010). *Nederland en de 'Responsibility to Protect': de verantwoordelijkheid om mensen te beschermen tegen massale wreedheden* (Vol. 70). The Hague: AIV.

Adviesraad Internationale Vraagstukken (AIV). (2012). *Europese Defensiesamenwerking. Soevereiniteit en handelingsvermogen* (Vol. 78). The Hague: AIV.

Adviesraad Internationale Vraagstukken (AIV). (2013). *Wisselwerking tussen actoren in internationale samenwerking* (Vol. 82). The Hague: AIV.

Adviesraad Internationale Vraagstukken (AIV). (2014). *De* EU-*gasafhankelijkheid van Rusland. Hoe een geïntegreerd* EU-*beleid dit kan verminderen* (Vol. 26). The Hague: AIV.

Adviesraad Internationale Vraagstukken (AIV). (2015). *Instabiliteit rond Europa: confrontatie met een nieuwe werkelijkheid* (Vol. 94). The Hague: AIV.

Adviesraad Internationale Vraagstukken (AIV). (2016). *Advies over veiligheid en stabiliteit in Noordelijk Afrika* (Vol. 101). The Hague: AIV.

Algemene, R. (2016). *Aandachtspunten bij de ontwerpbegroting 2017 (Chapter X) van het Ministerie van Defensie*. Algemene Rekenkamer: The Hague.

Amersfoort, H. (2016). Nederland, de weg kwijt. Over de teloorgang van de militaire strategie en de noodzaak van geschiedenis. *Militaire. Spectator, 185*(5), 217–231.

Analistennetwerk Nationale veiligheid. (2016, November). *Nationaal Veiligheidsprofiel 2016. Een All Hazard overzicht van potentiële rampen en dreigingen die onze samenleving kunnen ontwrichten*. Bilthoven: RIVM.

Andersson, J., Biscop, S., Giegerich, B., Molling, C., & Tardy, T. (2016). *Envisioning European defence. Five futures*. Paris: EU Institute for Security Studies (Chaillot paper).

E. Hirsch Ballin et al. (eds.), *Security in an Interconnected World*, Research for Policy, https://doi.org/10.1007/978-3-030-37606-2

Andrews-Speed, P., Bleischwitz, R., Boersma, T., Johnson, C., Kemp, G., & VandeVeer, S. D. (2012). The global resource nexus. The struggle for land, energy, food. In *Water and minerals*. Washington, DC: Transatlantic Academy.

Anthony, I. (2014). Introduction. International security, armaments and disarmament. In SIPRI *yearbook 2014. armament, disarmament and international security* (p. 1). Oxford: Oxford University Press.

Antonio Missiroli (ed.), *Towards an eu global strategy. Background, process, references*, European Union Institute for Security Studies, 2015.

Archer, C., Bailes, A., & Wivel, A. (Eds.). (2014). *Small states and international security. Europe and beyond*. London: Routledge.

Bailes, A. J. (2013). Societal security and small states. In C. Archer, A. J. K. Bailes, & A. Wivel (Eds.), *Small states and international security* (pp. 66–78). New York: Routledge.

Bailes, A. J., & Sando, C. (2014). *Nordic cooperation on civil security: The 'Haga' process, 2009–2014* (Occasional paper). Stockholm/Reykjavik: FOI/Centre for Small State Studies. http://ams.hi.is/wp-content/uploads/2014/04/The-Haga-Process-pdf.pdf

Bailes, A., Rickli, J.-M., & Thorhallsson, B. (2013). Small states, survival and strategy. In C. Archer, A. J. K. Bailes, & A. Wivel (Eds.), *Small states and international security* (pp. 26–44). New York: Routledge.

Bakker, A., & Drent, M. (2016). *Meerjarige Defensie Akkoorden in Nederland*. Instituut Clingendael: The Hague.

Bakker, A., Drent, M., Landman, L., & Zandee, D. (2016). *A stronger CSDP: Deepening defence cooperation*. Clingendael: The Hague.

Ballester, B. (2013). *The cost of Non-Europe in common security and defence policy*. Brussels: European Parliamentary Research Service.

Balzacq, T., Guzzini, S., Williams, M. C., Waver, O., & Patomaki, H. (2015). What kind of theory – If any – Is securitization? *International Relations, 29*(1), 96–136.

Beck, U. (1992). *Risk society; towards a new modernity*. London: Sage.

Beeres, R., & Bakker, E.-J. (2010). Drie misvattingen over bezuinigen bij Defensie. *Internationale Spectator, 64*(12), 657–659.

Beken, V. (2011). Grensvervaging tussen interne en externe veiligheid. *Tijdschrift voor Veiligheid, 10*(4), 59–69.

Bekkers, F., & de Spiegeleire, S. (2010). Strategische lenigheid. Anticperen op een fundamenteel onzekere toekomst. *Militaire Spectator, 179*(7/8), 360–371.

Biscop, S. (2015). *Peace without money, war without Americans. Can European strategy cope?* Farnham: Ashgate.

Bogdanor, V. (Ed.). (2005). *Joined-up government*. Oxford: Oxford University Press.

Bortslap, H. (2002). *Kracht zonder macht. Einddocument Werkgroep Strategische Functie op Rijksniveau.*

Bourbeau, P. (Ed.). (2015). *Security - Dialogue across disciplines*. Cambridge: Cambridge University Press.

Boutellier J. C. J. (2007). *Nodale orde. Veiligheid en burgerschap in een netwerksamenleving*. Oratie Vrije Universiteit. Amsterdam: VU.

Brattberg, E., & Hamilton, D. S. (2014). Recommendations. In E. Brattberg & D. S. Hamilton (Eds.), *Global flow security. A new security agenda for the transatlantic community in 2030*. Washington, DC: Center for Transatlantic Relations.

Brocades Zaalberg, T. W. (2008). The roots of Dutch Counterinsurgency. Balancing and integrating military and civilian efforts from Aceh to Uruzgan. *The U.S. army and irregular warfare 1775–2007*. Selected papers from the 2007 conference of Army Historians, Washington, DC.

Buijze, G. (2013, November 6–9). Grondstoffen inzet politiek spel. *Evo Logistiek*.

Buzan, B. (1991). *People, states and fear: An agenda for international security studies in the post-cold war era* (second ed.). London: Harvester Wheatsheaf.

Buzan, B., Waver, O., & de Wilde, J. (1998). *Security: A new framework for analysis*. Boulder: Lynne Rienner.

Canada, Privy Council Office. (2004). *Securing an open society: Canada's National Security Policy*.
Canadian GAO. (2015). *Fiscal sustainability of the armed force*.
Casteleijn, L. C. C. (2014). De trends van twintig jaar defensie; de motie Van der Staaij in historisch perspectief. *Atlantisch perspectief, 8*, 9–12.
Casteleijn, L. (2015). De collectieve verdediging en de Nederlandse krijgsmacht. *Internationale Spectator, 69*, 3.
Castells, M. (1996). *The rise of the network society*. Oxford: Blackwell.
Castells, M. (2009). *Communication power*. Oxford: Oxford University Press.
Center for Strategic & International Studies. (2016). *The kremlin playbook. Understanding Russian influence in central and Eastern Europe*. Washington, DC: CSIS.
Centre for European Policy Studies. (2015). *More union in European defence*. Brussels: CEPS.
Chakrabarti, S. (2014). Corporate actors. In M. Kaldor & Rangelov (Eds.), *The handbook of global security* (pp. 505–524). London: Wiley.
Christensen, T., & Laegreid, P. (2007). Whole of government approach, regulation, performance, and public sector reform. *Public Administration Review, 67*(6), 1059–1066.
Christou, G. (2014). The European Union's human security discourse: Where are we now? *European Security, 23*(3), 364–381.
Clark, C. (2013). *The sleepwalkers: How Europe went to war in 1914*. London: Penguin.
Clingendael. (2013). *Clingendael's visie op de krijgsmacht van de toekomst*. Instituut Clingendael: The Hague.
Clingendael. (2017). *Strategische monitor 2017*. Clingendael: The Hague.
Coker, C. (2014). *Globalisation and insecurity in the twenty-first century: NATO and the management of risk*. London: Routledge.
Colijn, K. (2016, Februari, 1). Een integraal veiligheidsbeleid is er nog altijd niet. *Vrij Nederland*.
Collinson, S., Duffield, M., Berger, C., da Costa, C.F., & Sandstrom, K. (2013). *Paradoxes of presence: Risk management and aid culture in challenging environments*. London: Humanitarian Policy Group, Overseas Development Institute.
Cooper, T., Metz, S., & Shah, M. (2011). *A more violent world? Global trends in organized violence*. Berghof Forschungszentrum fur Konstruktive Konfliktbearbeitung.
Corry, O. (2010). *'Securitization and "Riskization": Two grammars of security' in 7th Pan-*European International Relations Conference, Stockholm.
Cottey, A. (2013). *Security in 21st century Europe* (2nd ed.). Houndmills: Palgrave Macmillan.
Crawford, J. (2012). Sovereignty as a legal value. In J. Crawford & M. Koskenniemi (Eds.), *The Cambridge companion to international law* (pp. 117–133). Cambridge: Cambridge University Press.
Daase, C. (2010). *Der erweiterte Sicherheitsbegriff* (Working Paper). Frankfurt: Goethe- Universität.
Daase, C. (2013). Von der nationalen zur menschlichen Sicherheit: politische und rechtliche Konsequenzen des erweiterten Sicherheidsbegriff. In A. Fischer- Lescano & P. Mayer (Eds.), *(2010) Recht und Politik globaler Sicherheit. Bestandaufnahme und Erklärungsansätze* (pp. 11–42). Campus Verlag: Frankfurt/New York.
de Bakker, E., & Beeres, R. (2013). Hoeveel investeren is duurzaam? *Militaire Spectator, 182*(12), 548–560.
de Carvalho, B., & Sande Lie, J. H. (2015). A great power performance: Norway, status and the policy of involvement. In B. de Carvalho & I. B. Neumann (Eds.), *Small state status seeking. Norway's quest for international standing* (pp. 56–72). New York: Routledge.
de Coning, C., & Friis, K. (2011). Coherence and coordination the limits of the comprehensive approach. *Journal of International Peacekeeping, 15*(1/2), 243–272.
de Graaf, B. (2010). The Dutch mission in Uruzgan. Political and military lessons. *Atlantisch Perspectief, 34*, 7.
de Graaf, B. (2012a). Waar is de onzichtbare vijand? *Militaire spectator, 181*(1), 14–24.
de Graaf, B. (2012b). Nationale veiligheid in historisch perspectief I. In Nationaal Coördinator Terrorismebestrijding en Veiligheid. *Magazine Nationale veiligheid en crisisbeheersing* (pp. 48–51), 10, 4.

de Graaf, B. (2012c). Nationale veiligheid in historisch perspectief II. In Nationaal Coördinator Terrorismebestrijding en Veiligheid. *Magazine Nationale veiligheid en crisisbeheersing* (pp. 32–35), 10, 5.

de Graaf, B. (2012d). Nationale veiligheid in historisch perspectief III. In Nationaal Coördinator Terrorismebestrijding en Veiligheid. *Magazine Nationale veiligheid en crisisbeheersing* (pp. 52–55), 10, 6.

de Graaf, B., Dimitriu, G., & Ringsmose, J. (Eds.). (2015). *Strategic narratives, Public opinion and war*. New York: Routledge.

de Ridder, M. (2017). Internationale effecten circulaire economie vragen om moreel en strategisch leiderschap. *Internationale Spectator, 2017*, 1.

de Spiegeleire, S., Sweijs, T., Wijninga, P., & van Esch, J. (2014). *Designing future stabilisation efforts*. The Hgaue: HCSS.

de Vries, S. (2014). Europese defensiesamenwerking: een exercitie in de kantlijn. *Atlantisch Perspectief, 38*, 1.

de Wijk, R. (2010). Balkenende's defensie: doorgaand verval met behoud van kwaliteit. *Internationale Spectator, 64*, 454–458.

de Wijk, R. (2014). Nederlandse politieke keuzes en de internationale gevolgen. In Raad voor Maatschappelijke Ontwikkeling (Ed.), *Nationaal belang in meervoud. Perspectieven op internationale politiek* (pp. 60–82). The Hague: Raad voor Maatschappelijke Ontwikkeling.

de Wijk, R. (2015a). *Machtspolitiek*. Amsterdam: Amsterdam University Press.

de Wijk, R. (2015b). Nationale veiligheid in een woelige wereld. In Nationaal Coördinator Terrorismebestrijding en Veiligheid. *Magazine Nationale veiligheid en crisisbeheersing* (pp. 52–55), 13(3), 34–35.

de Wijk, R., & Toxopeus, R. (2005). Hoe binnen- en buitenlandse veiligheid verweven zijn. *Internationale Spectator, 59*(7/8), 421–425.

de Wilde, J. (2008). Speaking or doing human security? In M. den Boer & J. de Wilde (Eds.), *The viability of human security* (pp. 225–225). Amsterdam: Amsterdam University Press.

Deitelhoff, N., & Fischer-Lescano, A. (2013). Politik und Recht der privatisierten Sicherheit. In A. Fischer-Lescano & P. Mayer (Eds.), *Recht und Politik globaler Sicherheit. Bestandaufnahme und Erklärungsansätze* (pp. 43–77). Campus Verlag: Frankfurt/New York.

den Boer, M., & de Wilde, J. (2003). Top-down and bottom-up approaches to human security. In M. den Boer & J. de Wilde (Eds.), *The Viability of Human security* (pp. 9–18). Amsterdam: Amsterdam University Press.

Department for International Development (DFID). (2014). *Operational plan 2011–2016*. Climate and Environmental Department.

der Verteidigung, B. (2016). *Weissbuch. Zur Sicherheitspolitik und zur Zukunft der Bundeswehr*. Berlin.

Dimitriu, G., & de Graaf, B. (2011). De missie in Uruzgan als strategisch narratief. *Atlantisch Perspectief*, p. 8. www.atlcom.nl/ap_archive/pdf/AP%202011%20nr.%208/Dimitriu%20en%20De%20Graaf.pdf

Drent, M., & Meijnders, M. (2015) *Multi-year Defence Agreements. A Model for Modern Defence?*. The Hague: Instituut Clingendael.

Drent, M., & Zandee, D. (2015). Samen uit, samen thuis. Internationale defensiesamenwerking in een veranderde wereld. *Internationale Spectator, 69*, 3.

Drent, M., & Zandee, D. (2016, November 3). EU security and defence at home and abroad: From strategy to action. *Internationale Spectator*. www.internationalespectator.nl/article/eusecurity-and-defence-home-and-abroad-strategy-action#social-shares

Drent, M., Landman, L., & Zandee, D. (2014). *The EU as a security provider*. Clingendael: The Hague.

Drent, M., Landman, L., & Zandee, D. (2016). *A new strategy. Implications for CSDP*. Clingendael: The Hague.

Dreyer, I., & Stang, G. (2013). Foresight in governments – Practices and trends around the world. *Yearbook of European Security*, 1–26.

Ducheine, P. A. L. (2008). Krijgsmacht, geweldgebruik en terreurbestrijding. In *Een onderzoek naar juridische aspecten van de rol van strijdkrachten bij de bestrijding van terrorisme*. Nijmegen: Wolf Legal Publishers.

Ducheine, P. (2016). Je hoeft geen zwaard en schild te dragen om ridder te zijn. Mythen over digitale oorlogvoering en recht. *Militaire Spectator, 185*(3), 123–135.

Duyvesteyn, I. G. B. M. (2013). *Strategisch analfabetisme. De kunst van strategisch denken in moderne militaire operaties*. Oratie. Leiden: Universiteit Leiden. https://openaccess.leidenuniv.nl/bitstream/handle/1887/20943/Oratie%20Duyvesteijn%20NL.pdf?sequence=2

Economist. (2015, Maart 7). *The new nuclear age*.

Edmunds, T. (2012). British civil–military relations and the problem of risk. *International Affairs, 88*(2), 265–282.

Elhefnawy, N. (2011). Twenty years after the cold war: A strategic survey. *Parameters, 41*(1, Spring), 6–17.

Engbersen, G. et al. (2015). *Geen tijd verliezen: van opvang naar integratie van asielmigranten* (WRR-Policy brief 4). The Hague: WRR.

Eriksson, J., & Rhinard, M. (2009). The internal—External security nexus. Notes on an emerging research agenda. *Cooperation and Conflict, 44*(3), 243–267.

Factsheet Versterking Civiel – Militaire Samenwerking (VCMS). www.infopuntveiligheid.nl/Publicatie/DossierItem/41/4853/vcms%2D%2D-factsheet-vcms.html

Farrell, T., & Schmitt, O. (2012). *The causes, character and conduct of armed conflict , and the effects on civilian populations, 1990–2010*. Geneve: UNHCR.

Ferdowsi, M. A. (Ed.). (2002). *Internationale Politik im 21. Jahrhundert*. Munchen: Wilhelm Fink Verlag.

Final report of Future Policy Survey 2010 – *A new foundation for the Netherlands Armed Forces* 276; 303, annex to Parliamentary Documents ii 2009–2010, 31,243, no. 17.

Franke, V. C., & Dorff, R. H. (2012). *Conflict management and "whole government": Useful tools for U.S. National Security Strategy?* Carlisle Barracks: SSI.

Freedman, L. (2013). *Strategy. A history*. Oxford: Oxford University Press.

Freedman, L. (2014). Ukraine and the art of crisis management. *Survival, 56*(3), 7–42.

French White Paper. (2013). *Defence and national security*.

Frerks, G., & Klein Goldewijk, B. (2007). Human security: Mapping the challenges. In G. Frerks & B. Klein Goldewijk (Eds.), *Human security and international insecurity* (pp. 21–44). Wageningen: Wageningen Academic Publishers.

Gehem, M., Usanov, A., Frinking, E., & Rademaker, M. (2015). *Assessing cyber security. A meta-analysis of threats, trends, and responses to cyber attacks*. The Hague: HCSS.

Ghemawat, P. (2014). *The Netherlands in Europe*. https://www.wrr.nl/publicaties/verslag/2014/04/16/verslag-seminar-4-nederland-en-de-europese-unies

Goede, P. J. M., Goverde, H. J. M., & Nelissen, N. J. M. (2001). De netwerksamenleving volgens Manuel Castells. *Bestuurswetenschappen, 55*(2), 101–136.

Graduate Institute Geneva. (2011). *Small arms survey 2011*. www.smallarmssurvey.org/?leadmin/docs/A-Yearbook/2011/en/Small-Arms-Survey-2011-Chapter-04-en.pdf

Grandia, M. (2015). *Deadly embrace? The decision paths to Uruzgan and Helmand*. Proefschrift Universiteit Leiden. https://openaccess.leidenuniv.nl/handle/1887/32625

Granholm, N., & Malminen, J. (2014). A strategic game changer? In N. Granholm, J. Malminen, & G. Persson (Eds.), *A rude awakening. Ramifications of Russian aggression towards Ukraine* (pp. 9–15). Stockholm: FOI.

Gray, C. (2010). *The strategy bridge. Theory for practice*. Oxford: Oxford University Press.

Gray, C. (2014). *Strategy and defence planning; meeting the challenge of uncertainty*. Oxford: Oxford University Press.

Griffith, I. L. (1996). *Caribbean security on the Eve of the 21st century*. Washington, DC: Institute for National Strategic Studies/National Defense University.

Griffiths, R. T. (2014). Economic security and size. In C. Archer, A. J. K. Bailes, & A. Wivel (Eds.), *Small States and International Security* (pp. 46–65). New York: Routledge.

Gruiters, J. (2011). ISAF-evaluatie. Bescheiden maar omkeerbare verbeteringen in Uruzgan en onzekere toekomst voor Afghanistan. *Atlantisch Perspectief,* p. 8. www.atlcom.nl/ap_archive/pdf/AP%202011%20nr.%208/Gruiters.pdf

Guzzini, S. (Ed.). (2012). *The return of geopolitics in Europe? Social mechanisms and foreign policy identity crises*. Cambridge: Cambridge University Press.

Habegger, B. (2009). *Strategic foresight: Anticipation and capacity to act*. Zurich: Center for Security Studies.

Hagmann, J. (2015). *(In-)security and the production of international relations. The politics of securitisation in Europe*. London/New York: Routledge.

Hague Centre for Strategic Studies, The (HCSS). (2011). *Op weg naar een grondstoffenstrategie. Quick scan ten behoeve van de Grondstoffennotitie*. The Hague: HCSS.

Hague Centre for Strategic Studies, The (HCSS). (2013). *De toekomst in alle staten*. HCSS *strategische monitor 2013*, The Hague: HCSS.

Hague Centre for Strategic Studies, The (HCSS). (2014). *Strategic monitor 2014. Four strategic challenges*. The Hague: HCSS.

Hague Centre for Strategic Studies, The (HCSS). (2015). *Strategic monitor 2015. The Return of Ghosts Hoped Past? Global Trends in Conflict and Cooperation*. The Hague: HCSS.

Hague Centre for Strategic Studies, The (HCSS). (2016). *The wheel of fortune. Strategic monitor 2016*. The Hague: HCSS.

Hague Centre for Strategic Studies, The (HCSS). (2017). *Een kompas voor een wereld in beweging. De rol van Buitenlandse Zaken in het borgen van Nederlandse belangen*. The Hague: HCSS.

Hague Centre for Strategic Studies, The (HCSS) & TNO. (2014a). *The Global resource Nexus. Impact on sustainable security of supply of agri-food imports for The Netherlands*. The Hague: HCSS & TNO.

Hague Centre for Strategic Studies, The (HCSS) & TNO. (2014b). *The geo-politics of shale gas*. The Hague: HCSS & TNO.

Hameiri, S., & Jones, L. (2013). The politics and governance of non_traditional security. *International Studies Quarterly, 57*(3), 462–473.

Hammerstad, A., & Boas, I. (2014). National security risks? Uncertainty, austerity and other logics of risk in the UK government's National Security Strategy. *Cooperation and Conflict*, december 2015, *50*(4), 475–449.

Hasenclever, A. (2014). Human dignity and war. In M. Duwell, J. Braarvig, et al. (Eds.), *The Cambridge handbook of human dignity: Interdisciplinary perspectives* (pp. 439–445). Cambridge: Cambridge University Press.

Hazelbag, L. (2015). Nationale Veiligheidsraad: politiek wenselijk en staatsrechtelijk haalbaar? *Militaire Spectator, 184*(4), 184–197.

Hellema, D. (2014). *Nederland in de wereld*. Utrecht: Spectrum.

Herington, J. (2015). Philosophy: The concepts of security, fear, liberty, and the state. In P. Bourbeau (Ed.), *Security - dialogue across disciplines* (pp. 22–44). Cambridge: Cambridge University Press.

Hirsch, Ballin, E. M. H. (2014). Ontwikkeling van de internationale rechtsorde. *Christen Democratische Verkenningen, 34*, (winter), 136–142.

Hoffenaar, J. (2009). *Een politieke aangelegenheid en daarmee nooit een uitgemaakte zaak. De ontwikkeling van de hoofdtaken en het ambitieniveau van de Nederlandse krijgsmacht na de Koude Oorlog*. Nederlands Instituut voor Militaire Historie.

Hoffman, F. G. (2009). Hybrid threaths: Reconceptualizing the evolving character of modern conflict. *Strategic Forum, 240*, 1–8.

Homan, C. (2005). *De krijgsman als ontwikkelingswerker?* www.clingendael.nl/sites/default/files/20051100_cscp_art_homan.pdf

Homan, C. (2006). Defensie en interne veiligheid: van vangnet naar volwaardige veiligheidspartner. *Armex, 90*(1), 21–24.

Homan, C. (2007). De militair en wederopbouw. *Internationale Spectator, 61*(2), 63–67.

Homan, C. (2008). Klimaatverandering als veiligheidsprobleem. *Atlantisch Perspectief, 32*, 12–17.
Homan, C. (2014). ODA en het Budget Internationale Veiligheid. *Internationale Spectator, 68*(7/8), 55–57.
Horton, A. (1999). A simple guide to successful foresight. *Foresight, 1*(1), 5–9.
Hough, P. (2013). *Understanding global security* (3rd ed.). New York: Routledge.
House of Commons Defence Committee. (2016). *Russia: Implications for UK defence and security*. London: House of Commons Library.
House of Commons, United Kingdom. (2010a). *The comprehensive approach: The point of war is not just to win but to make a better peace* (Seventh report of session 2009–10). www.publications.parliament.uk/pa/cm200910/cmselect/cmdfence/224/224.pdf
House of Commons, United Kingdom. (2010b). Public administration select committee. Who does UK National strategy? www.publications.parliament.uk/pa/cm201011/cmselect/cmpubadm/435/435.pdf
House of Lords, European Union Committee, United Kingdom. (2015). *The EU and Russia: Before and beyond the crisis in Ukraine*. London: House of Lords.
Human security Research Group (HSRG). (2005). *Human security report 2005: War and peace in the 21st century*. Vancouver: Simon Fraser University.
Human security Study Group. (2016). From hybrid peace to human security: Rethinking EU strategy towards conflict. www.securityintranistion.org
Hutchison, E., & Bleiker, R. (2014). Theorizing emotions in world politics. *International Theory, 6*(3), 491–515.
Inspectie Ontwikkelingssamenwerking en Beleidsevaluatie (IOB). (2013). *Investeren in stabiliteit. Het Nederlandse fragiele statenbeleid doorgelicht*. The Hague: IOB.
Institute for Public Policy Research (IPPR). (2009). *Shared responsibilities. A National Security Strategy for the UK, the final report of the IPPR commission on National Security in the 21st century*. London: IPPR.
International Bank for Reconstruction and Development (IBRD). (2011). *World development report 2011. Conflict , security and development*. Washington, DC: The World Bank.
International Dialogue on Peacebuilding & Statebuilding. (2014). *New deal monitoring report 2014*.
International Organization for Migration (IOM). (2008). *Migration and climate change*. Geneve: IOM.
Jentleson, B. W. (2009). An integrated executive branch strategy. In D. W. Drezner (Ed.), *Avoiding Trivia; the role of strategic planning in American Foreign Policy* (pp. 60–83). Washington, DC: Brookings.
Jermalavicius, T., Pernik, P., & Hurt, M. (2014). *Comprehensive security and integrated defence: Challenges of implementing whole of government and whole of society approaches*. Tallin: International Centre for Defence Studies. http://curis.ku.dk/ws/files/99129162/icds_Report_Comprehensive_Security_and_Integrated_Defence_10_February_2014.pdf
Judt, T. (2005). *Postwar. A History of Europe since 1945*. New York: Penguin Books.
Jung, D. W. G., & Molenaar, K. (2013). 'Nuchter omgaan met risico's. Bewust omgaan met veiligheid. *Magazine Nationale Veiligheid en Crisisbeheersing, 11*(6), 38–41.
Kaldor, M. (1998). *New and old wars*. Cambridge: Cambridge University Press.
Kaldor, M. (2014). Human security. In M. Kaldor & I. Rangelov (Eds.), *The handbook of global security*. London: Wiley.
Kaldor, M., & Rangelov, I. (Eds.). (2014). *The handbook of global security*. London: Wiley.
Kasselmann, H. J. (2012). Civil-military cooperation. A way to resolve complex crisis situations. *PRism, 4*(1), 17–29.
Kissinger, H. A. (1957). *Nuclear power and foreign policy*. New York City: Harper/Oxford: Oxford, hier geciteerd naar de verkorte heruitgave, New York: Norton 1969.
Kitzen, M. (2008). Westerse militaire cultuur en counter-insurgence. Een tegenstrijdige realiteit. *Militaire Spectator, 177*(3), 123–134.

Klein Goldewijk, B., & Frerks, G. (2007). Human security and international security: The quest for human global governance. In G. Frerks & B. Klein Goldewijk (Eds.), *Human security and international insecurity* (pp. 265–295). Wageningen: Wageningen Academic Publishers.

Klem, B., & Frerks, G. E. (2004). *Evaluatie Vredesfonds : Flexibiliteit met visie 2001–2003*. The Hague: Instituut Clingendael. http://edepot.wur.nl/41301.

Klem, B., & Frerks, G. E. (2007). *Evaluatie Stabiliteitsfonds 2004–2005*. www.bartklemresearch.nl/files/publications/evaluatie_stabiliteitsfonds_def.pdf

Klep, C. (2012). Mission Uruzgan. *Atlantisch Perspectief,* p. 7. www.atlcom.nl/ap_archive/pdf/AP%202012%20nr.%207/Klep.pdf

Korteweg, A. R. (2011). *The superpower, the bridge-builder and the hesitant ally. How defense transformation divided NATO (1991–2008)*. Leiden: Leiden University Press.

Kramer, F. D. (2014). Irregular conflict and the wicked problem dilemma. Strategies of imperfection. *PRism, 2*(3), 75–100. http://cco.ndu.edu/Portals/96/Documents/prism/prism_2-3/Prism_75-100_Kramer.pdf.

Krepinevich, A. F., & Watts, B. D. (2009). *Regaining strategic competence. Strategy for the long haul*. Washington, DC: Center for Strategic and Budgetary Assessments.

Krom, A., van Waes, A., van Est, R., & Brom, F. (Eds.). (2015). *Grondstoffenhonger duurzaam stillen*. The Hague: Rathenau Instituut.

Larsen, K., & Håkenstad, M. (2012). *Long-term defence planning: A comparative study of seven countries* (p. 5). Oslo: Oslo Files on Defence and Security.

Le Billion, P. (2015). Geography: Securing places and spaces of securitization. In P. Bourbeau (Ed.), *Security – Dialogue across disciplines* (pp. 62–89). Cambridge: Cambridge University Press.

Leidraad Geïntegreerde Benadering. (2014). *De Nederlandse visie op een samenhangende inzet op veiligheid en stabiliteit in fragiele staten en conflict gebieden*. The Hague.

Lindley-French, J., & Tjepkema, A. (2010). *Between the polder and a hard place? The Netherlands and Defence planning challenges for smaller European countries*. London: RUSI.

Liotta, P. H., & Owen, T. (2006). Why human security? *Whitehead Journal of Diplomacy and International Relations, 7*(37), 37–54.

Losman, D. (2001). Economic security a national security folly? *Policy Analysis, 409*, 1–12.

Luizard, P.-J. (2015). *Le piège Daech ; L'État islamique ou le retour de l'histoire*. Paris: La Decouverte.

Lynn, B. C. (2014). Shock therapy: Building resilient international industrial systems in 2030. In E. Brattberg & D. S. Hamilto (Eds.), *Global flow security. A new security agenda for the transatlantic community in 2030* (pp. 185–205). Washington, DC: Center for Transatlantic Relations.

Malik, S. (2015a). Constructing security. In P. Hough et al. (Eds.), *International security studies. Theory and practice* (pp. 72–85). London: Routledge.

Malik, S. (2015b). Human security. In P. Hough et al. (Eds.), *International security studies. Theory and practice* (pp. 57–72). London: Routledge.

Martin, M., Kaldor, M., & Serro, N. (2013). Introduction. In *National, European and human security. From co-existence to convergence* (pp. 1–17). New York: Routledge.

Martinage, R. (2015). Under the sea. The vulnerability of the commons. *Foreign Affairs, 94*, 117.

Matthies, V. (2002). Krisenpravention und Friedenskonsolidierung. In M. A. Ferdowsi (Ed.), *Internationale Politik im 21. Jahrhundert* (pp. 123–143). Wilhelm Fink Verlag: München.

Matthijsen, C. J. (2014). Van 3D naar geïntegreerde benadering. Een beeld van de ontwikkelingen sinds Uruzgan. *Militaire Spectator, 183*(5), 228–239.

Meessen, W., & van der Wiel, W. (2014). Concept Development & Experimentation (CD&E). *Militaire Spectator, 183*(4), 206–216.

Ministerie van Buitenlandse Zaken. (2013). *Internationale Veiligheidsstrategie. Veilige Wereld, Veilig Nederland*. Den Haag: Ministerie BZ.

Ministerie van Buitenlandse Zaken. (2014). *Beleidsbrief Internationale Veiligheid – Turbulente tijden in een instabiele omgeving*. The Hague: Ministerie van Buitenlandse Zaken.

Ministerie van Buitenlandse Zaken. (2016). *Planetary security: Peace and cooperation in times of climate change and global environmental challenges*. The Hague: Ministerie van Buitenlandse Zaken (Conference report, 2–3 November 2015, Vredespaleis, Den Haag).

Ministerie van Binnenlandse Zaken en Koninkrijksrelaties. (2007). *Strategie Nationale Veiligheid*. Den Haag: Ministerie BZK.

Ministerie van Defensie. (2017). *Houvast in een onzekere wereld. Lijnen van ontwikkeling in het meerjarig perspectief voor een duurzaam gerede en snel inzetbare krijgsmacht*. Ministerie van Defensie: The Hague.

Ministerie van Financiën. (2015). *Meer 'Bang for the Buck'. Interdepartementaal beleidsonderzoek naar meer doelmatigheid door internationale samenwerking en geïntegreerde contracten bij wapensystemen van Defensie*. The Hague.

Ministry of Defence, United Kingdom. (2014). *Global strategic trends – Out to 2045*.

Missiroli, A. et al. (2014, December). A changing global environment. *Chaillot papers*.

Moïsi, D. (2009). *The Geopolitics of Emotions: How Cultures of Fear, Humiliation, and Hope are Reshaping the World*. New York: Random House.

Moïsi, D. (2009/2015). *De geopolitiek van emotie. Over angst, vernedering en hoop, en de opgave voor deze tijd*. Amsterdam: Nieuw Amsterdam.

Munich Security Conference. (2017). *Munich security report 2017 Post-truth, Post-west, Post-order?* Munich security conference.

Murray, W. (2001). *Thinking about innovation, Naval War College Review*. Spring, liv, 2.

Nationaal Coördinator Terrorismebestrijding en Veiligheid (NCTV). (2014). *Magazine Nationale Veiligheid en Crisisbeheersing*, 12, 1.

Nationaal Cyber Security Centrum. (2016). *Cybersecuritybeeld Nederland 2016*. The Hague: Ministerie van Justitie.

National Security Coordination Centre. (2004). *The fight against terror. Singapore's National security strategy*. Singapore. www.nscs.gov.sg/public/download.ashx?id=48

Odgaard, L. (2014). *Strategy in NATO: Preparing for an imperfect world*. London: Palgrave Macmillan.

Olmer, J. (2016). *Globale Migration: Geschichte und Gegenwart (Beck'sche Reihe)*. München: Beck.

Organisation for Economic Cooperation and Development (OECD). (2003). *Emerging risks in the 21st century – An agenda for action*. Paris: OECD.

Organisation for Economic Cooperation and Development (OECD). (2006). *Whole of government approaches to fragile states*. Paris: OECD.

Organisation for Economic Cooperation and Development (OECD). (2012). *Estonia: Towards a single government approach*. Paris: OECD.

Organisation for Economic Cooperation and Development (OECD). (2013). *OECD risk management. Strategic crisis management*. Paris, OECD.

Organisation for Economic Cooperation and Development (OECD). (2015). *States of fragility 2015. Meeting Post-2015 ambitions*. Paris: OECD.

Osterhammel, J. (2011). *Die Verwandlung der Welt: eine Geschichte des 19. Jahrhunderts*. München: Beck.

Pai, M. A., & Khan, M. A. (1974). On-line security assessment and control of interconnected power systems using sensitivity concept. In *4th* IFAC/IFIP *International Conference on Digital Computer Applications to Process Control* (pp. 304–316). Heidelberg/Berlin: Springer.

Palm, T. (2016). Consequenties van een Brexit voor het Europees veiligheidsbeleid. *Internationale Spectator*, 3.

Pelkmans, J. (1982). The many faces of national economic security. In F. A. M. Alting von Gesau & J. Pelkmans (Eds.), *National economic security: Perceptions, threats and policies* (pp. 3–16). Tilburg: John F. Kennedy Institute.

Persson, G., & Pallin, C. V. (2014). Setting the scene – The view from Russia. In N. Granholm, J. Malminen, & G. Persson (Eds.), *A Rude awakening. Ramifications of Russian aggression towards Ukraine* (pp. 25–34). Stockholm: FOI.

Pilbeam, B. (2015a). New wars, globalisation and failed states. In P. Hough et al. (Eds.), *International security studies. Theory and practice* (pp. 104–119). London: Routledge.
Pilbeam, B. (2015b). The rise of private military and security companies. In P. Hough et al. (Eds.), *International security studies. Theory and practice* (pp. 191–209). London: Routledge.
Pingeot, L., & Obenland, W. (2014). *In whose name? A critical view on the responsibility to protect*. New York: Rosa Luxemburg Stiftung.
Pinker, S. (2011). *The better angels of our nature. The decline of violence in history and its causes*. Penguin UK.
Planbureau voor de Leefomgeving (PBL). (2014). *Balans van de leefomgeving* 2014. *De toekomst is nú*. The Hague: PBL.
Power, M. (2004). The risk management of everything. *The Journal of Risk Finance, 5*(3), 58–65.
Prevost, G., Vanden, H. E., Campos, C. O., & Ayerbe, L. F. (Eds.). (2014). *US National security concerns in Latin America and the Caribbean*. New York: Palgrave Macmillan.
Princen, F., & Wijnen, M. (2004). Defensieplanning. *Militaire Spectator, 173*(12), 616–628.
Putten, F.-P., Rood, J., & Meijnders, M. (2016). *Grootmachten en mondiale stabiliteit. Clingendael Monitor 2016*. The Hague: Instituut Clingendael.
Rand Corporation. (2016). *Reinforcing deterrence on nato's eastern flank. Wargaming the defense of the Baltics*. Rand: Arroyo Center.
Rapportage Werkgroep Economische Veiligheid. (2014). *Tussen naïviteit en paranoia. Nationale veiligheidsbelangen bij buitenlandse overnames en investeringen in vitale sectoren*. The Hague: Ministerie van Veiligheid en Justitie.
Rathenau Instituut. (2017). *Een nooit gelopen race. Over cyberdreigingen en versterking van weerbaarheid*. The Hague: Rathenau Instituut.
Raynova, D., & Kearns, I. (2015). *The Wales pledge revisited: A preliminary analysis of 2015 budget decisions in NATO member states*. London: European Leadership Network.
Realuyo, C. B. (2013). *Collaborating to Combat Illicit networks through interagency and international efforts*. Washington, DC: Center Hemisphere Defense Studies, National Defense University. http://chds.dodlive.mil/files/2013/12/pub-OP-realuyo.pdf
Rekenkamer, A. (2012, December 11). *Onderzoek 'budgettaire gevolgen van de beleidsvoornemens over internationaal klimaatbeleid en internationale veiligheid voor ontwikkelingssamenwerking'*.
Rekenkamer, A. (2014, oktober 7). Aandachtspunten bij de begroting 2015 van het Ministerie van Economische Zaken.The Hague: Algemene Rekenkamer.
Rekenkamer, A. (2015a). *Bezuinigingen en intensiveringen bij de* AIVD. *Gevolgen van de budgettaire turbulentie in de periode 2012–2015*. The Hague: Algemene Rekenkamer.
Rekenkamer, A. (2015b). *Stimulering van duurzame energieproductie (*SDE+*) Haalbaarheid en betaalbaarheid van beleidsdoelen*. The Hague: Algemene Rekenkamer.
Renn, O. (2008). *Risk governance. Coping with uncertainty in a complex world*. New York: Earthscan.
Ries, T. (2014). Global flow security: A conceptual framework. In E. Brattberg & D. S. Hamilton (Eds.), *Global flow security. A new security agenda for the transatlantic community in 2030* (pp. 3–22). Washington, DC: Center for Transatlantic Relations.
Rood, J. (2010). Nederlands buitenlands beleid in de spiegel van de toekomst. *Internationale Spectator, 64*(1), 8–11.
Rood, J. (Ed.). (2014). *Een wankele wereldorde. Clingendael Strategische Monitor 2014*. The Hague: Clingendael.
Rood, J., & Dinissen, R. (Eds.). (2013). *Een wereld in onzekerheid. Clingendael Strategische Monitor 2013*. The Hague: Clingendael.
Rood, J., van der Putten, F.-P., & Meijnders, M. (2015). *Een wereld zonder orde. Clingendael Strategische Monitor 2015*. The Hague: Clingendael.
Rosenberg, E. S. (Ed.). (2012). *A world connecting, 1870–1945*. Cambridge, MA: Belknap Press of Harvard University Press.

Schoenmaker, B. (2012). *Vredesoperaties en de terugkeer van de oorlog, oratie*. Leiden: Universiteit Leiden.
Schout, A., & Rood, J. (2013). Europa voorbij de beeldvorming. In A. Schout & J. Rood (Eds.), *Nederland als Europese lidstaat: eindelijk normaal?* (pp. 15–30). Den Haag: Boom Lemma.
Schramade, K. (2016). *De kaalslag bij Defensie. Amsterdam*. Amsterdam: University Press.
Selchow, S. (2014). Security policy and (global) risk(s). In M. Kaldor & I. Rangelow (Eds.), *The handbook of global security policy* (pp. 68–84). Hoboken: Wiley Blackwell.
Seybolt, T. B. (2008). *Humanitarian military intervention. The conditions for success and failure*. New York: Oxford University Press/SIPRI.
Shell. (2013). *New lens scenario's. A shift in perspective for a world in transition*. The Hague: Shell International BV.
Simma, B., & Müller, A. T. (2012). Exercise and limits of jursidiction. In J. Crawford & M. Koskenniemi (Eds.), *The Cambridge companion to international law* (pp. 134–157). Cambridge, Cambridge University Press.
Stehfest, E., Vuuren, D., van Kram, T., & Bouwman, L. (2014). *Integrated assessment of global environmental change with image. Model description and policy applications*. The Hague: Planbureau voor de Leefomgeving.
Stiftung Wissenschaft und Politik (SWP) and German Marshall Fund of the United States (GMF). (2013). *New power. New responsibility. Elements of a German foreign and security policy for a changing world*. Berlin/Washington, DC: SWP and GMF.
Stojkovic, D., & Dahl, B. R. (2007). *Methodology for long term defence planning*. Norwegian Defence Research Establishment (FFI).
Stolberg, A. G. (2012). *How nation-states craft national security documents*. Strategic Studies Institute, U.S. Army War College. www.strategicstudiesinstitute.army.mil/pubs/display.cfm?pubID=1128
ter Beek, R. (1996). *Manoeuvreren. Herinneringen aan Plein 4*. Amsterdam: Uitgeverij Balans.
ter Haar, B., & Maas, E. (2014). *Nederlandse percepties van bedreigingen en uitdagingen. Samenvatting en analyse van een onderzoek voor de OVSE*. The Hague: Instituut Clingendael.
Terpstra, N., Willems, R., Frerks, G., & Chang Pico, T. (2014). What the New deal can learn from the Human security approach, Research report. *Knowledge Platform Security and Rule of Law*.
U.S. Marine Corps. (2015). 2015 Marine corps security environment forecast. Futures 2030–2045. Quantico.
United National Development Programme. (2014). *Human development report 2014*.
United Nations Development Programme. (1994). *Human development report 1994*. New York/Oxford: Oxford University Press.
van der Lijn, J. (2011). *'3D the next generation'. Lessons learned from Uruzgan for future operations*. The Hague: Instituut Clingendael.
van der Lijn, J., Briscoe, I., Drent, M., Homan, K., van der Putten, F.-P., & Zandee, D. (2015). *Peacekeeping operations in a changing world*. The Hague: Instituut Clingendael.
van Eekelen, W. F. (2013). Nederland – Europa of Amerika. In A. Schout & J. Rood (Eds.), *Nederland als Europese lidstaat: eindelijk normaal?* (pp. 67–81). The Hague: Boom Lemma.
van Esch, J., de Jong, S., & de Ridder, M. (2014). *No blood for oil? Economic security, energy security and the military*. The Hague: HCSS.
van Haaster, J., & Roorda, M. (2016). The impact of hybrid warfare on traditional operational rationale. *Militaire Spectator, 185*(4), 175–185.
van Schaik, L., Maas, E., Dinissen, R., & Vos, J. (2015). *Beyond scares and tales: Climate- Proofing Dutch Foreign Policy* (Clingendael report for the Netherlands Environmental Assessment Agency). The Hague: Clingendael.
van Staden, F. (2013). De EU: International speller met gebreken. In A. Schout & J. Rood (Eds.), *Nederland als Europese lidstaat: eindelijk normaal?* (pp. 49–66). Boom Lemma: The Hague.
van Veen, E., & Dudouet, V. (2017). *Hitting the target, but missing the point? Assessing donor support for inclusive and legitimate politics in fragile societies*, OECD International Network on Conflict and Fragility.

Vlek, C. (2013). How solid is the Dutch (and British) National risk assessment? Overview and decision theoretic evaluation. *Risk Analysis, 33*(6), 948–971.

Voorhoeve, J. J. C. (1979). *Peace, profits and principles: A study of Dutch foreign policy*. The Hague: Martinus Nijhoff.

Voorhoeve, J. J. C. (2010). Van medespeler tot figurant? *Internationale Spectator, 64*(4), 189–190.

Wagen, C. M. (2012). *Twenty-first century defense and disruptive innovation*. Carlisle Barracks: US Army War College.

Wallensteen, P. (2012). *Understanding conflict resolution: War, peace and the global system*. London: Sage.

Waver, O. (1995). Securitization and desecuritization. In R. D. Lipschutz (Ed.), *On security*. New York: Columbia University Press.

Weidenhaus, G. (2015). *Soziale Raumzeit*. Berlin: Suhrkamp.

Weller, E. (2014). *Comprehensive approach to human security*. Research report. *Knowledge Platform Security and Rule of Law*.

Werner, W. (2015). International law: Between legalism and securitization. In P. Bourbeau (Ed.), *Security - Dialogue across disciplines* (pp. 196–218). Cambridge: Cambridge University Press.

Wetenschappelijke Raad voor het Regeringsbeleid (WRR). (2001). *Ontwikkelingsbeleid en goed bestuur*. The Hague: Sdu.

Wetenschappelijke Raad voor het Regeringsbeleid (WRR). (2006). *Dynamiek in islamitisch activisme. Aanknopingspunten voor democratisering en mensenrechten*. Amsterdam: Amsterdam University Press. https://english.wrr.nl/publications/reports/2006/04/18/dynamism-in-islamic-activism

Wetenschappelijke Raad voor het Regeringsbeleid (WRR). (2007). *Europa in Nederland* [Rediscovering Europe in the Netherlands]. Amsterdam: Amsterdam University Press. https://english.wrr.nl/publications/reports/2007/06/05/rediscovering-europe-in-the-netherlands

Wetenschappelijke Raad voor het Regeringsbeleid (WRR). (2008). *Onzekere veiligheid* [Uncertain Safety]. Amsterdam: Amsterdam University Press. https://english.wrr.nl/publications/reports/2008/10/01/uncertain-safety

Wetenschappelijke Raad voor het Regeringsbeleid (WRR). (2010a). *Minder pretentie, meer ambitie. Ontwikkelingshulp die verschil maakt* [Less pretension, more ambition. Development policy in times of Globalization]. Amsterdam: Amsterdam University Press. https://english.wrr.nl/publications/reports/2010/01/18/less-pretention-more-ambition

Wetenschappelijke Raad voor het Regeringsbeleid (WRR). (2010b). *Aan het buitenland gehecht. Over verankering en strategie van Nederlands buitenlands beleid[Attached to the World*: On the anchoring and strategy of Dutch foreign policy]. Amsterdam: Amsterdam University Press. https://english.wrr.nl/publications/reports/2010/11/30/attached-to-the-world

Wetenschappelijke Raad voor het Regeringsbeleid (WRR). (2011). *Evenwichtskunst. Over de verdeling van verantwoordelijkheid voor fysieke veiligheid [Physical Safety*. A Matter of Balancing Responsibilites], Amsterdam: Amsterdam University Press. https://english.wrr.nl/publications/publications/2012/10/09/physical-safety

Wetenschappelijke Raad voor het Regeringsbeleid (WRR). (2013). *Naar een lerende economie. Investeren in het verdienvermogen van Nederland*, Amsterdam: Amsterdam University Press. https://www.wrr.nl/publicaties/rapporten/2013/11/04/naar-een-lerende-economie

Wetenschappelijke Raad voor het Regeringsbeleid (WRR). (2014). *Consistent maatwerk – handreikingen voor dossieroverstijgend risico- en veiligheidsbeleid*. The Hague: WRR. https://www.wrr.nl/publicaties/brieven/2014/06/26/consistent-maatwerk%2D%2D-handreikingen-voor-dossieroverstijgend-risico%2D%2Den-veiligheidsbeleid

Wetenschappelijke Raad voor het Regeringsbeleid (WRR). (2015). *De publieke kern van het internet. Naar een buitenlands internetbeleid*. [The Public Core of the Internet. An International agenda for Internet Governance]. Amsterdam: Amsterdam University Press. https://english.wrr.nl/publications/reports/2015/10/01/the-public-core-of-the-internet

Weterings, R., Bastein, T., Tukker, A., Rademaker, M., & de Ridder, M. (2013). *Rousources for our future. Key issues and best practices in resource efficiency*. Amsterdam: HCSS, TNO & Amsterdam University Press.

Wiers, J. (2016). *Mind the Gap. Hoe de kloof tussen buitenlandbeleid en wetenschap te overbruggen?, oratie*. Groningen: Rijksuniversiteit.

Wijnbelt, M., Klijn, H., Faber-Bosma, J., van Bergenhenegouwen, J., & Bangoer, A. (2013). *Een ingebeeld kwaal. Het strategisch vermogen van de Rijksoverheid*. The Hague: NSOB.

Williams, M. J. (2009). *NATO, security and risk management: From Kosovo to Khandahar*. Oxon: Routledge.

Winkler, H. A. (2015). *Geschichte des Westens: die Zeit der Gegenwart*. München: Beck.

Wittkowsky, A., & Wittkampf, U. (2013, Januari, 1–4). Pioneering the comprehensive approach: How Germany's partners do it. *Policy Briefing*. Zif Center for International Peace Operations.

Wivel, A., Bailes, A. J. K., & Archer, C. (2014). Setting the scene: Small states and international security. In C. Archer, A. J. K. Bailes, & A. Wivel (Eds.), *Small states and international security* (pp. 3–24). New York: Routledge.

Wolf, M. (2014a). Shaping globalization. *Finance and Development, 51*, 3. www.imf.org/external/pubs/ft/fandd/2014/09/wolf.htm

Wolf, M. (2014b). Shaping Globalization. *Finance and Development, 41*(3), 1–4.

World Bank. (2014). *Word development report 2014. Risk and opportunity. Managing risk for development*. Washington, DC: World Bank.

World Economic Forum. (2013). *Global risks 2013* (8th edn.). Geneva: World Economic Forum. http://reports.weforum.org/global-risks-2013/

World Economic Forum. (2015). *Global risks 2015* (10th edn.). Geneva: World Economic Forum. http://reports.weforum.org/global-risks-2015/

Zakaria, F. (1997, November/December). The rise of liberal democracy. *Foreign Affairs*.

Zandee, D. (2016, November 2). Defensie onder Hennis: niet meer dan een herstelkrijgsmacht. *Internationale Spectator*. www.internationalespectator.nl/article/defensie-onder-hennis-niet-meerdan-een-herstelkrijgsmacht#_ednref13

Zandee, D., Drent, M., & Hendriks, R. (2016). *Defence cooperation models. Lessons learned and usability*. The Hague: Clingendael Institute.

Zarkow, D., & Hintjens, H. (2015). Introduction. Conflict , peace, security and development: Theories and methodologies. In H. Hintjens & D. Zarkov (Eds.), *Conflict , peace, security and development. Theories and methodologies* (pp. 1–21). New York: Routledge.

The manufacturer's authorised representative in the EU is Springer Nature Customer Service Centre GmbH, Europaplatz 3, 69115 Heidelberg, Germany. If you have any concerns regarding our products, please contact ProductSafety@springernature.com

Printed and bound by CPI Group (UK) Ltd, Croydon, CR0 4YY
30/09/2026
02225610-0006